RADIO'S NICHE MARKETING REVOLUTION

Broadcasting & Cable Series

Series Editor: Donald V. West, Editor/Senior Vice President, *Broadcasting & Cable*

Global Television: How to Create Effective Television for the 1990s
Tony Verna

The Broadcast Century: A Biography of American Broadcasting
Robert L. Hilliard and Michael C. Keith

Global Telecommunications: The Technology, Administration, and Policies
Raymond Akwule

Selling Radio Direct
Michael C. Keith

Electronic Media Ratings
Karen Buzzard

International Television Co-Production: From Access to Success
Carla Brooks Johnston

Practical Radio Promotions
Ted E.F. Roberts

The Remaking of Radio
Vincent M. Ditingo

Winning the Global Television News Game
Carla Brooks Johnston

Merchandise Licensing for the Television Industry
Karen Raugust

Radio Programming: Tactics and Strategies
Eric Norberg

Broadcast Indecency: FCC Regulation and the First Amendment
Jeremy Harris Lipschultz

Copyrights and Trademarks for Media Professionals
Arnold P. Lutzker

Radio's Niche Marketing Revolution: FutureSell

Godfrey W. Herweg
Ashley Page Herweg

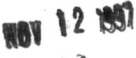

Focal Press
Boston Oxford Johannesburg Melbourne New Delhi Singapore

Focal Press is an imprint of Butterworth–Heinemann.

Copyright © 1997 by Butterworth–Heinemann
ℜ A member of the Reed Elsevier group
All rights reserved.

No part of this publication may be reproduced, stored in a retrieval system, or transmitted in any form or by any means, electronic, mechanical, photocopying, recording, or otherwise, without the prior written permission of the publisher.

∞ Recognizing the importance of preserving what has been written, Butterworth–Heinemann prints its books on acid-free paper whenever possible.

Butterworth–Heinemann supports the efforts of American Forests and the Global ReLeaf program in its campaign for the betterment of trees, forests, and our environment.

Library of Congress Cataloging-in-Publication Data
Herweg, Godfrey W.
  Radio's niche marketing revolution : FutureSell / Godfrey W. Herweg, Ashley Page Herweg.
     p.   cm.
  Includes bibliographical references and index.
  ISBN 0-240-80202-0 (alk. paper)
    1. Radio advertising—United States. 2. Market segmentation—United States.   I. Herweg, Ashley Page.   II. Title.
HF6146.R3H39   1997
659.14'2--dc21   97-16724
                                                                      CIP

British Library Cataloguing-in-Publication Data
A catalogue record for this book is available from the British Library.

The publisher offers special discounts on bulk orders of this book.
For information, please contact:

Manager of Special Sales
Butterworth–Heinemann
225 Wildwood Avenue
Woburn, MA 01801
Tel: 617-928-2500
Fax: 617-928-2620

For information on all Focal Press publications available, contact our World Wide Web home page at:
http://www.bh.com/bh

10 9 8 7 6 5 4 3 2 1
Printed in the United States of America

Contents

Preface   xi
Acknowledgments   xiii

Part One: Marketing Options

1

The Niche Marketing Revolution 3
What Is Niche Marketing? 3
Cracks in the Melting Pot 4
A Mass Requiem 5
All Kinds of Niches 5
Customer Needs 8
The Tip of the Advertising Iceberg 8
Can You Afford Niche Research? 9
FutureMedia 10
The Information Revolution 10
Next 11

2

Didn't Radio Invent Niche Marketing? 12
Not Really 12
Combo Thinking 12
Mass Marketing's Era 13
The Arrow 14
Interactive—Not for Video Alone 15
New Technology and Niche Marketing 16
Are Radio's Negative Factors Positives? 17
How Small Is Your Niche? 19
Being Different 20

Part Two: Research

3

Using Niche Research to Increase Listenership 25
Great Sums of Time and Money 25
Using Station Information 25
Database Mountains 27
Information Retrieval 34
The Difference Is Listening 34

4

Using Niche Research to Increase Sales 35
Sleuthing 35
Fast Forward to the Present 38
Share-of-Advertiser 40
Step One 41
Congratulations 53

5

Database Marketing 54
Learning the Tush-Push 54
Shrinking Media Budgets 54
Mountains of Information 55
Interactive Options 57
Database Wars 59
The Price Trap 59
Database Commitments 60
Advantages of Database Marketing 62
A Fresh Start 67

6

Focusing on Focus Groups 68
Should You Consider Focus Group Research? 68
Commissioning the Study 70
Focus Group Composition 72
The Need for Ongoing Research 74
Ensuring Successful Research Results 78
Starting and Conducting the Interview 80
After the Interview 83
Power 85

7

Listening for Niches 86
Simplifying Life 86
The Early Days and Today's Salespeople 87
The Cost of Business 88
A New Approach for a New Era 89
Ask Buyers 93
Is This Necessary? 94

Part Three: People

8

Niche Marketing Managers—A New Breed 99
Manager Number Five 99
What Niche Managers Need to Know 100
A New Breed 102
Opening the Secret Diary—Start with the Billing Printout 108
The Big Picture 111
Inner Peace 115
Managing 115
The Journey 116

9

Niche Marketing Sales Consultants—Their Difference *Is* Knowledge 117
What Holds People Back? 117
Effective Training for Niche Consultants 117
Giving Encouragement 131
Advertisers as Friends and Family 131
Preparing to Create Wealth 132
A Sitcom 133

Part Four: Presentations and Niche Copywriting

10

A Niche Presentation for an Auto Dealer 137
Great Books Versus Great Presentations 137
The Assignment 137
Reality 153

11

A Niche Presentation for a Building Center 155
Insert Your Format 155
Mike, The Futurist 165

12

Tabulating Survey Results 166
The First Step 166
The Survey Is Completed—Help! 166
Making Sense Step-by-Step 167
Presenting Your Findings 172

13

Presenting Survey Results 173
Searching for Niches 173
Looking Forward 181

14

Introducing Niche Marketing Services to Advertisers 182
A Niche Marketing Presentation 182
Afterword 191

15

Copywriting for Niches 193
The Importance of Questioning 193
Learn to Know Thy Consumer 193
Advertising Strategy 202
Testing Appeals 205
Making Your Job Easier 205
Copywriter Sleuthing 206
Becoming a Good Self-Critic 213
Who Has THE Answer? 213

Part Five: Clients as Partners and Looking Ahead

16

Partnering with Ad Agencies 217
Advertising Agents and Agencies 217
Media Buyers a.k.a. Negotiators 221

The Media Planner Needs Analysis   224
Challenges, Creativity, and Technology   229
Agencies of the Future   231
What to Expect From Agencies and Advertisers   232

17

The Importance of Building Partnerships   234
A Dream Come True   234
Trouble in Paradise   236
Delivering More Than You Promised   241
Analyzing Your Best   242
Being a Good Partner   244
Nurturing Partnerships   246
Crystallizing Expectations   251
Is It Easy to Do Business with Your Radio Station?   254

18

The Future   255
Groupies   255
Is There a Future for Individual Stations?   256
FutureMedia   257
Research Is Key   257
Another Perspective   258
The Odds Are in Your Favor   259
The New Breed of Radio Salespeople   259

References   261

Suggested Further Reading   263

After Words   265

About the Authors   267

Index   269

Preface

Radio's Niche Marketing Revolution: FutureSell is about the mercurial changes radio stations face as local markets continue to fragment. You can cope—and grow—in this volatile marketing climate. We'll show you how.

The choices are simple. You can cut rates and engage in deadly price wars. *Or*, you can change the way you do business. Traditionally, the focus has been on programming; that is, fighting for increased audience shares. Instead, every department can begin to focus on marketing. This means learning to understand your advertisers' needs, not from your point of view, but from their point of view and then, from their customers' point of view.

In the era of mass marketing, short-term profits were a primary focus of broadcasters. The results of this short-term thinking have been numerous takeovers, mergers, and consolidations. Also, stations are "going dark" in record numbers. You can stop this radio-station graveyards explosion. At the same time, you can strengthen the new multistation groups, which are forming in single markets. Long-term planning and niche marketing strategies provide some of the tools. We'll show you why.

Niche marketing research is dramatically different from mass-marketing research. Gross rating points, cumes, and average-quarter-hour estimates are now obsolete measurements. The quality of your individual research and the quality of your advertisers' results will provide the new standards of measurement. A sample in-house database is offered. You'll see advertiser presentations that have been street-tested and developed to gain solid, niche marketing results. Presenting the *quality* of your listeners, instead of the *quantity* of your listeners, will help separate your station from the long line of other media options. The emphasis is no longer on mass audiences . . . rather it is on *niche* audiences, which can produce advertising results. We'll give you concrete examples.

The rise and fall of written and face-to-face presentation techniques has been fostered in part by dwelling mainly on numbers and fax machines. Knowing as much—or more—than your competitors about *all* media helps you gain business. The presentations in Chapters 10 through 14 show you how to create and deliver niche marketing presentations that can lead to success in fragmenting markets. You'll see how radio is being used more effectively. Niche copywriting is discussed to help you understand the power radio can enjoy even though copy and production changes are frequent.

Niche marketing is a great opportunity for radio broadcasters, ad agencies, and advertisers. Niche marketing provides a path to new revenue streams. The niche marketing parade never stops. Change is the norm. To succeed, its imperative for *everybody* on your staff to meet customers and then, listen and learn their needs.

The focus is no longer on chasing larger audience shares. Small is beautiful—and more profitable—in the new era of niche marketing.

The ideas you'll get for *your* radio station as you read this book are the most valuable gifts you'll take away from it. Get a pad of paper and a pen, or get ready to write in the margins. Sit back, relax your shoulders, and enjoy this glimpse into the future of radio.

Acknowledgments

Two authors are listed on the cover of this book. As anyone who has ever written a book can tell you, "it takes a village" of caring, interested, and supportive people to help make your *seed thought* a reality. Even though Raymond A.L. Herweg is no longer here with us, a portion of his enormous wealth of knowledge has been passed to us. He could be called the third author of this book. Thank you, Ray, for being a person whose thirst for knowledge was never quenched. At the time of his passing—at 92 years— he was teaching himself to speak German—his *sixth* self-taught language. You've given this family a standard for which to strive—for generations.

Thank you Alison Garrett, personal friend and editor-extraordinaire, for giving so much of your precious time to edit and make insightful, constructive comments about our manuscript. The jokes included in the margins and the motivational notes at the end of each chapter were precious words. And, thanks for those Monday-morning faxes—it was a great way to start the week.

The following people have made invaluable suggestions for improving this book:

> Anthony Smith, captain of the support team; Gary Fries, President/CEO, Radio Advertising Bureau; Rhody Bosley, Partner, Research Director, Inc.; Skip Finley, CEO/COO, American Urban Radio Network; Rick Padulo, President/CEO, Padulo Integrated, Inc.; Bob Jordan, President, and Jim Higginbotham, Chairman, The Media Audit; Edie Jett, Sherry Stevens, Beth McClellan, Janice Wilson, Tina Messinger, Marshall Simon, Crystal Branham, and Chuck Herndon.

Some of them have allowed us to use their real-life stories in order to help others learn through their experiences. We thank them for their unique perspectives and wisdom.

And, finally, thanks to our inspiration . . . Chelsea, Hope, Geoff, Jonathan, and Michael.

PART ONE
Marketing Options

1

The Niche Marketing Revolution

What Is Niche Marketing?

Niche marketing may be broadly described as the process of finding goods and services which absolutely delight specifically defined customers in markets that are increasingly fragmented. Once you discover what delights these customers, you try to increase customer loyalty; you're no longer trying to get as many customers as possible. You're concentrating on increasing purchases from the customers you have while you try to find more like them.

Incredible Opportunities

Radio stations, advertising agencies, and traditional mass marketers like Coca-Cola® are beginning to see the incredible opportunities niche marketing offers. This book is designed to help you find more riches in profitable niches. The ideas you are about to encounter work for multinational media conglomerates, stations in small markets, and duopolies in any size market.

Is Niche Marketing for Me?

Market fragmentation, which spawns niche markets, is happening all over the globe. Niche marketing is an outstanding growth strategy for media. Radio companies and advertisers who understand and embrace niching will have a competitive edge in the twenty-first century.

The Numbers Fix

Traditionally, mass media has found that the fastest way to increase billing is by increasing their audience numbers. This approach appeals to many media buyers as well. Advertisers take comfort in the fact that station "A" has more 25- to 54-year-old numbers than station "B." Media-buying services use mass numbers, or the lack of them, to drive station rates down. Account executives can cover weak media decisions by pointing to *the numbers*. As media options increase, however, mass audience numbers will continue to decline. In ads on the cover of Standard Rate and Data Service (SRDS), radio giants like WCCO in Minneapolis once boasted that, "WCCO Radio had more audience than all the TV stations in Minneapolis combined." Not so today—or tomorrow. Now, however, there are riches in niches.

Cracks in the Melting Pot

Historically, North America has been perceived as a mass market. Waves of immigrants came to the United States and Canada. Communication with their cultural past was limited. As a result, the collective immigrants were assimilated into mainstream society. North America was seen as a melting pot for cultural differences. Today, however, our melting pot is cracking. Thanks, in part, to worldwide communications and transportation, cultural diversity is maintained and intact. Immigrants and minorities in North America, Western Europe, and soon the majority of our planet, can easily stay in touch with their ethnic, cultural, and religious roots. Radio, television, telephones, computer networks, and jumbo jets make communicating easier. The paradox is that these massive communications pipelines are now fueling market fragmentation.

The Illusion of National Oneness

Today, the mass market—like the melting pot—is history. Our ancestors successfully promoted the idea of one nation. Advertising and marketing people have spent billions of dollars in the United States, millions of deutsche marks in Germany, and millions of yen in Japan appealing to the illusion of national oneness. The melting pot in Russia, Europe, Asia, and North America, however, has segmented into millions of mercurial pools. Today it's quite feasible for an immigrant from India, who drives a taxi in Atlanta or London, to return home for a visit on a yearly basis. A Philippine desk clerk at a Virginia Holiday Inn may maintain her cultural roots by returning to her homeland for a three-month holiday. The ability of these new immigrants to retain strong cultural ties with their homelands is part of the reason we're entering the new era of niche marketing.

Jerry Falwell and the Pope

U.S. broadcasters have the opportunity to cater to much, much more than a black-and-white North America with Western European tastes. For example, the United States has had an influx of Asian immigrants who represent hundreds of subcultures. There's also a multitude of Hispanics with a mosaic of different tastes and information needs. An astute marketer will recognize that sometimes subcultures will merge briefly to form special interest groups. For example, the Roman Catholic church may form an alliance with Christian fundamentalists to oppose gay rights. The issue may be as simple as criticizing a furniture store ad that targets gay couples.

A Mass Requiem

The smokestack economies created by the nineteenth-century Industrial Revolution thrived on mass marketing. For nearly a hundred years, media and advertisers have genuflected at the altar of mass numbers. Media has tried to deliver boxcar numbers to satisfy advertisers selling goods and services. Advertising appeals were usually to the lowest common denominator—the mass market—or an unrealistic demo such as adults 25–54. Twenty-five-year-olds have very different consumer tastes than 54-year-olds. Smaller demographic cells like Southern Chinese women 25 to 34 with disposable incomes of $18,000 make more sense. You scoff? In northern California, this is a valuable niche.

The Homogeneous Market Myth

To date, only 8% of companies in the developed nations are actively engaged in niche marketing. These blue-chip companies include some impressive names such as IBM, Saab, Hewlett-Packard, Office Depot, American Express, Red Lobster, British Airways, Holiday Inns, Jones New York, Opti-World, Minolta, and American Airlines. Our studies show that the 8% figure is even lower for media companies. We predict, however, that in the first decade of the twenty-first century more than 80% of companies, including the media, will be actively engaged in some form of niche marketing.

All Kinds of Niches

Yesterday

In the early 1940s, there was only one version of Coca-Cola. It arrived in a 6-ounce green bottle (Figure 1.1). The only difference between a bottle of Coke sold in Toronto and a bottle of Coke sold in Santa Fe was the name of the bottler on the bottom of the container.

Today

The actions of former mass-marketing leaders like Coca-Cola are worth noting. Coke built a soft-drink empire on a single product—a carbonated drink made with caffeine, caramel coloring, and sugar. In the early 1970s, the company de-emphasized a successfully niched diet drink, Tab, because the sales figures weren't large enough—even though Tab was profitable. Today, Coke is diversifying. They're experimenting with a myriad of soft

6 Marketing Options

FIGURE 1.1 A 6-ounce bottle of Coke.

drinks to compete with Snapple, Lipton Iced Teas, and others. For example, in the early 1990s, some Coca-Cola bottlers test-marketed a soft drink called Fruitopia, which is made with noncarbonated mineral water and natural fruit juices. The market segment may be tiny compared to Coke Classic. Profitable niches, however, offer growth opportunities for this former mass-marketing giant. Coke has experimented and enjoyed some success with Royal Mystic—a soft drink with a natural fruit juice base and spring water. Coca-Cola test-marketed OK—a similar soft drink product aimed at the 18–24-year-old male niche.

Tomorrow

Why did the Coca-Cola company change its views about smaller market segments? The diverse tastes of consumers create the need for niched products. Coca-Cola test-markets many flavors; through sampling and promotion, Coca-Cola determines which flavors are currently in greatest demand. They expect that flavor demands will be different in different parts of the country. The Coke company also expects that flavor demands may change with the seasons—creating a series of different niches. This represents a great opportunity for radio stations to become part of the sampling and promotion process to help uncover niche markets and to research the bottlers' customer base.

McDonalds Joins the Parade

McDonalds is another example of a mass-marketing giant that has turned to niches. You can buy beer in a McDonalds in Spain or Germany. You can buy beer and wine in a McDonalds in Paris or Rome. You can buy meat pies in Australian McDonalds. In India, where the cow is sacred, you can buy bean-curd and veggie burgers. During some months of the year, you can buy chili at a McDonalds in Ladson, South Carolina. Then, add the new expanded *adult* menu at McDonalds. So much for the universal fare of McDonalds' hamburgers, french fries, and a Coke.

Niching for Moms

Hallmark Greeting Cards now markets over 1,200 different cards just for Mother's Day. Mom can receive greetings from spouses, parents, stepparents, children, stepchildren, and pets. Mother's Day greetings can now be given to mom-to-be, an aunt who's like a mom, a stepmom, a stepmom-in-law, a grandmom, even a great-grandmom. In today's marketplace, *Mom* has many niches.

A High-Flying Niche

Airlines have discovered that approximately 5% of the adult population make approximately 70% of ALL air trips. To court these lucrative customers, frequent-flyer programs have been created. A closer look at the frequent-flyer group reveals that there's a microgroup of people who take trips to places like Tokyo, Beijing, and New Delhi once a month, or more. To make sure the airlines can "romance" these important customers, programs like Delta's Million Miler Club were developed for people who fly a million miles or more per year. You can use the Delta example to explain the need for niching to your clients; large profits come from targeting small groups of a company's heavy users.

Why Florists Don't Advertise at Easter

Easter is the busiest season of the year for florists. Why not advertise? Many florists will tell you they don't need to advertise at Easter because they'll have a lot of business anyway. Often, florists feel that when they advertise an Easter Special, they are just advertising for the florist closest to the person hearing the ad. But, if you look more closely at their customer base, you'll realize that you don't need to wait for holidays to approach florists.

Floral Niches. A majority of florists' customers buy flowers for special occasions. For example, a customer may purchase flowers for her mom's birthday. If enough customers purchase flowers for a relative's birthday, you've uncovered a niche for your florist client. But don't write an ad for special-occasion giving. Write an ad targeting businessmen (if that's your demo) who send flowers to their mothers for their birthdays. Write a separate ad for each reason your audience will purchase flowers. Eliminating holidays, these ads may give you enough niches to develop a 52-week campaign!

Customer Needs

Many ad agencies are becoming aware of niche marketing. Agency executives are now seeking new ways to learn about the smallest needs of consumers. They're also beginning to cater to the perceptions of their clients' customers. For example, the Japanese have combined robotics and computer technology to create an ultimate niche marketing scenario. In Japan, when a consumer walks into a customized bicycle shop, a salesperson takes detailed notes about that consumer's needs. They perform a Customer Needs Analysis. The salesperson will ask, "Over what kind of terrain will you be riding? How far will you usually travel while you're riding your bicycle? How do you want it to feel underneath you?" Then, the salesperson measures the customer's body. The information is fed into a computer. The store's computer is on-line with the factory's computer. The customer's specifications reach the factory in nanoseconds, and, in less than 48 hours, the custom bicycle is ready. To establish value, however, the company takes two weeks to deliver the final product. Some people do not seem to associate speed with value—yet.

The Tip of the Advertising Iceberg

Thanks to computer scanning, customized clothing may be one of the next niche categories. With the use of six cameras, your body is completely scanned. Measurements are relayed to the manufacturer. Patterns are cut from the fabric with the use of a laser. Picture the perfect fit of every garment you purchase made to fit only your body. For future reference, your measurements are stored on a Smart Card that can be read at a retail location or at the manufacturer's location. Travel down to your foot. Once your feet are scanned, you may carry your measurements into a shoe store that carries *no* inventory. Customized shoes are ordered for your feet. Imagine the advertising opportunities created by the few examples here. You're experiencing the tip of a niche-marketing iceberg loaded with profitable opportunities for radio stations, advertisers, and ad agencies.

Venerable giants like U.S. Steel, IBM, and General Motors have been slow to change. These companies help us see the dilemma mass-market thinking creates. The mass marketers' response to change often has been to close plants, shrink their labor force, look for cheap labor overseas, and create their product line from a cookie cutter in an attempt to control costs and remain competitive. There are, however, viable alternatives, including looking for profitable niches in fragmenting markets. Will cars customized for your body type be the next niche? Some manufacturers are beginning to realize that styles and tastes are changing so fast that cheap overseas labor pools cannot respond quickly enough to meet new customer demands. One of the challenges of niche marketing is to stay abreast of fluid market situations.

What Can You Do?
Radio stations and ad agencies that understand these new niche opportunities can help their clients profit from increased market fragmentation. For example, you can offer to interview the local General Motors dealer's customers. You can ask them why they buy from your dealer—or why they don't. You'll find detailed survey forms for an automobile dealer in Chapter 10.

The Experience Factor
Ad agencies and broadcasters, or should we say *niche*casters, are beginning to understand that less expensive, but unskilled, account executives are not able to compete in niche markets. Niche markets may be created in an instant. The need for consumer research is immediate. To help clients compete, you can offer to perform frequent studies such as mall-intercept interviews and point-of-purchase surveys (i.e., when they make long-term commitments to your station). The information you glean will allow you to make superior client presentations as *niche* marketing consultants instead of being cast in a long line with other media salespeople.

Establishing Value
After interviewing prospective advertisers, a proficient sales consultant may be able to create an advertising presentation in just a few hours. To establish value, however, allow a week between the interview and the presentation. Examples of successful interviewing forms—Niche Marketing Analysis: Part I and Niche Marketing Analysis: Part II—can be found in Chapter 4 and Chapter 7, respectively.

Can You Afford Niche Research?

The question may be best phrased as, "Can you afford not to do niche research?" Consider your world. You live in the Information Age. Knowledge and the use of knowledge determine market leaders. Your company's success depends on the superior knowledge of your entire staff. New studies are needed to define niche audiences. The research will happen in real-time frames and won't rely primarily on the recollections of listeners or projected audience estimates. You may be researching your listeners—one listener at a time. You may also help your advertisers research their customers—one customer at a time. Can you afford this research? When it means more dedicated listeners and increased billing, the answer is simple: YES.

FutureMedia

When describing the media of the future, management guru Peter Drucker and futurist Alvin Toffler rarely mention radio. To change these perceptions of reality, it is necessary to repackage radio as a *new* medium. One way to repackage radio is to emphasize its newly refined, interactive ability. Engineers tell us that by the year 2010 every market on the surface of the earth could have at least 500 media options available. Radio can be imaged as the portable medium of choice. In this new media scenario, radio stations that grow will look for ways to *super*-serve the needs of their core audience and core advertisers, continually looking for new ways to present information. You'll ask your most loyal listeners—often one listener at a time—which kinds of information and entertainment interests them most. You'll learn to *re*invent radio based on information from the end-user, the listener.

The Information Revolution

In the age of information, with computers, satellites, and communication superhighways, some advertisers, like the Parker Pen Company, have tried to promote their goods using the same appeal around the globe. The net result, according to the *Wall Street Journal*, is that Parker has sold its pen division. Mass appeals—on an international, national, regional, or local market basis—just don't work well any more. Mass advertising appeals are less effective because people in individual cultures cling to their own set of values—creating new niche opportunities everywhere you look.

My Market Is Different

In the 1970s and 1980s, national radio representatives in many places, such as New York City, Chicago, Toronto, and Los Angeles, would hear a variation on the same theme from their client stations. Invariably, station management would insist that,"My market is different." The national radio rep would be invited to visit the market and would be given a tour of the local shopping malls and, perhaps, the tower site. Nothing much looked different in those days. The client would complain that prospects like Wal-Mart or Pizza Hut were not buying radio time. The standard answer was that radio did not generate enough 25–54 numbers. The buy went to television, newspapers, or preprinted inserts. The markets really didn't look very different from the rep's point of view. There seemed to be one seamless group of shopping malls selling the same variety of merchandise.

Market Differences

Today, you see that those markets are indeed different. In fact, giants, such as Campbell Soup, have created eight regional offices in the United States just to stay on top of regional and ethnic differences. Campbell's Tomato Soup may be more spicy in some regions of the country based on regional tastes uncovered by niche research. Thanks to massive immigration to other places, such as Germany, England, and the United States, ethnic differences are a reality in most of the world. Add the fact that, with today's instant communication systems, immigrants' ability to stay in touch with their "mother" mores, folkways, and culture is easy. Cultural differences are thriving. You're standing on the edge of a marketing revolution. The rules are changing as you read. Radio stations, ad agencies, and advertisers can reap huge profits from new niches that emerge in what seems like an instant.

Next

In the chapters that follow, you'll explore some of the formatting, technological, and positioning opportunities radio offers in niche formats. You'll see how daily radio, which is increasingly flexible, is a prime medium in the niche marketing revolution that is taking place—right now.

2

Didn't Radio Invent Niche Marketing?

Not Really

Magazines, direct mail, and cable television didn't invent niche marketing either. These media, however, with targeted formats, may look like they invented niche marketing. The niche marketing process, as you'll see, is far more complex than offering a specific format, a specialty magazine, or a special-interest cable channel. For example, let's look at a magazine for golfers: Is it directed toward mostly male golfers or female golfers? Does the editorial content target beginners or seasoned professionals? Are the readers retired players or recent college graduates? The considerations for possible niches grows continually.

Combo Thinking

Multiple group-owned stations, in markets like Boston, format to what may soon be considered broad demographics (e.g., 18–34). The multistation combinations in single markets are really an extension of mass-market thinking. In their book *Making Niche Marketing Work*, Robert Linneman and John Stanton (1994) point out that effectively reaching niche markets with existing media is difficult. Radio is in a unique position, however, to change that perception and deliver lucrative niches.

Unmarried Women

By creating multistation groups in single markets, some radio station owners believe that they are doing the same kind of niching as Coke, McDonalds, Delta Air Lines, and Hallmark. These multi-group stations offer a variety of what they believe are narrow-based demographics. Offering advertisers a radio station with a format that targets 18–34-year-olds in combination with another radio station that targets 35–44-year-olds is not niching. Instead, you'll need to use your programming and marketing microscopes. A niche specification will sound something like: unmarried women, 18–24, living in affluent city blocks with incomes of $40,000 plus. For a soft-drink client, for example, you'll also need to identify what percentage of this group likes soft drinks made with mineral water and natural fruit-flavored bases. Similar lifestyles—psychographics (i.e., consumers' interests and attitudes)—combined with very narrow geo-demographics (i.e., households and their composition within a specific geographic area) will become blueprints for defining marketing niches.

Mass Measurements Are Obsolete

The head-count mentality of newspaper circulation figures and television gross rating points are going down the same drain. Cost-per-point pricing and other mass-marketing measurements, which embrace 6- and 12-plus audience numbers, make little sense in the new era of niche marketing. Instead, you'll look for the *quality* of the audience in a daypart rather than the total number of listeners. You'll see more reasons why mass numbers no longer make sense in Chapters 3 and 4.

Mass Marketing's Era

Fifty Percent Off All Shoes

In the era of mass marketing, most retailers appealed to several market segments with one offer. For example, a shoe store would hold a store-wide sale using ads that say, "50% off all shoes. We have shoes for men, women, children, teens, and senior citizens." But niche marketing requires a sharper understanding of your market. Using traditional marketing methods, you split the shoe store's mass market into five segments. Niche marketing, however, requires an examination of far smaller segments than men, women, children, teens, and seniors. You'll need to take a close look at each of those five segments. For example, niching requires finding out why your teen customers would buy shoes from this store in the first place. Niching is not targeting persons 25–54 or even a demographic of men 25–34. Defining your niche may require one-to-one marketing. You'll need to ask teens individually why they shop at the shoe store. You'll learn about these techniques in Chapter 4.

Two Questions

In the new era of niche marketing, a question nichecasters may want to ask themselves is "Do the products or services we're advertising delight our core audience?" A second consideration may be "Will this niche be profitable for our clients and our radio station?"

Uncovering Rich Niches

Meet with the Mercedes Benz dealer. Learn all you can about the benefits of Mercedes Benz ownership. Then, ask the dealer's permission to interview his customers. Learn why his customers buy from him. At this point, you should have enough information to develop several multi-week campaigns. Now, you're niching for your clients.

Yes and No

If a nichecaster owns three stations in a market, should the stations offer very different formats to give the salespeople a broader base of advertisers with which to work? Should you offer an 18–24 demo on station one, a 25–34 demo on station two, and a 35–44 demo on station three? Looking through niche marketing eyes, the answer is, Yes, and No. Yes, if you research each format using the niche marketing techniques described in Chapter 3. No, if you're just playing hits from the sixties, seventies, and eighties and you think that's enough targeting to attract a loyal niche audience.

Narrowcasting—Not Yet!

There are successful all-sports, all-talk, all-news, shock-jock radio, and various ethnic formats such as urban and Hispanic. These broadbased radio stations may become niched. You have yet to see, however, the full impact of fragmentation on radio formats. You have seen some incredible declines in share points over the past two decades. For example, Duncan's *American Radio* reported a 31 share for WCCO in 1975. This giant station, with a 50,000 watt, clear-channel signal in Minneapolis, Minnesota, declined to an average 15 share in 1994. That's a 50% loss and the end is not in sight. In the marketplace, WCCO is not an isolated example.

The Arrow

As audiences shrink, radio programmers continue to look for format alternatives. In Los Angeles, most current niches are filled—two or three times over. Looking for a niche, KCBS introduced "The Arrow." Because this niche format only plays songs from the 1970s, the question may be asked, "Is The Arrow narrow enough?" Perhaps it's too broadbased for Los Angeles, Tokyo, Rome, or Moscow. Perhaps The Arrow fills a sufficient niche in Edmonton, Alberta; Lyon, France; or Waco, Texas.

A Lesson From Sondra Gair

Once upon a time there were radio dramas like *Sky King* and *The Lone Ranger*. During the 1940s, Sondra Gair was a radio actor who starred in those programs. Like many other old-time actors, she had a voice so well trained that she could make the recitation of numbers in a phone book sound interesting. She also understood niche opportunities. In 1986, Sondra had a brilliant idea; she wanted to create an international radio newscast—in Chicago. International news? In the United States? Only a senior citizen would be interested. "Preposterous!" said many radio executives. However, WBEZ—a public radio station—gave Sondra a chance. Her show's international coverage was pioneering and revolutionary. Sondra called newsmakers around the world and brought them to Chicago. Her program went off the air in March 1994, just before her death. Sondra's program filled a valuable programming niche in a major radio market. When you begin to examine your market's possible niches more closely, you may discover an exciting niche that no broadcaster is serving.

Many Urban Niches

A very successful radio niche is black radio. There are approximately 400 U.S. radio stations targeting African American listeners. There are also 40 different formats available to these 400 stations. Jay Williams, president of the American Urban Network in New York City, predicts 40 formats are just the beginning of the urban programming explosion. Owners of urban-format radio stations have a wide-open opportunity to uncover valuable marketing niches.

Interactive—Not for Video Alone

Radio has always been interactive. Live personalities answering the phone on-air have interacted with listeners since KDKA first started broadcasting in Philadelphia in 1920. Interactive media, like radio, has instant, real-time niche marketing research capabilities. Interactive radio and computers offer tremendous niche marketing potential. You can increase the effectiveness of an ad campaign by offering to combine direct-mail and radio. If you have a telephone sales department, you may want to consider using your telemarketing people to sell your advertisers' products and services as well. This could mean a whole new revenue stream for radio and added sales for your advertisers. Our research indicates that successful media in the twenty-first century will be interconnected as well as interactive.

FM Lessons From Long Ago

In the early 1950s, when AM radio was king, struggling FM stations offered low-cost conversion kits for AM car radios. Generally, the AM operators who enjoyed boxcar numbers laughed at their fledgling FM competitors. How times have changed. You can accomplish the same marketing strategy today with interactive radios placed in automobiles. Then, radio will be in an even stronger position to take the lead in the business of niche marketing.

Interactive Radio Technology Exists—Right Now!

There is no reason why radio can't take the lead in the interactive media business. According to Lee Simmons, a nationally recognized broadcast engineer, "Off-the-shelf technology exists for two-way radio today. Specialized Mobilized Radios (SMRs) offer the technology for two-way radio. There's room in the cellular band at 800 and 900 megahertz for an interactive system to operate." According to Simmons, "Building an interactive radio would be inexpensive. And, it's very easy to do." In fact, when somebody sets up an interactive television system, station operators can apply for a group of kilocycles for interactive radio. Mini-keyboards, similar to hand-held calculators, can be fitted to interactive radios for instant niching opportunities.

New Technology and Niche Marketing

Buy Cellular

In an era of niche marketing, you may find it profitable to buy a cellular phone frequency rather than another radio station in the same market. The interactive possibilities for profit are limitless. In-depth information from services—traffic, weather, and stockbrokers—can easily be accessed using cellular phones. The technology exists that allows people in their vehicles to dial a particular service using their cellular phone. Individuals can listen to headlines about a subject, such as pharmaceutical stocks, on their cellular phones; using a preassigned PIN code, they could get the latest quotes about specific stocks. Radio stations can advertise the stock services on-air. Then, the station could become a source of more in-depth information than one would regularly broadcast. Imagine being able to offer detailed information about an individual's stock portfolio. If you determine that your core audience invests heavily in stocks, this approach may make economic sense in the niche marketing era.

Win–Win Advertising Opportunities

The technology exists to make the instant translation scenario work. The stations, the advertiser, and the listener win using these innovations. You can quickly become connected with new niche markets when radio stations help market radios with instant translation systems. Voice-activated technology provides a connection to lucrative groups of new markets. Instant translations and other innovations can be sold by stations to help the new technology gain a foothold in local markets. The private sector now has an opportunity to fill some of the niches heretofore reserved for organizations such as the British Broadcasting Corporation (BBC) and Voice of America (VOA).

The Ag Example

Just about a hundred years ago nearly 95% of all people in North America were engaged in the business of agriculture. Or they were in a business that supported the agricultural industry. Today, the figure stands at approximately 3% and, as other nations enter the Information Age, their agricultural populations will also decline. The agricultural market in Europe and the Americas was once a massive market driven by the brute force of millions of people. This is an excellent example of a mass market that is now niched. The agricultural market shrank because of the explosion of knowledge coupled with computers and other new technologies. Technology is a contributing factor in the creation of niche markets.

Cable Radio

In the early 1980s, we, the authors, took an option on a cable radio channel. We owned a mass-market AM station with a signal covering three distinct areas: portions of Ontario, Quebec, and the United States. Research showed a U.S. niche—a number of alumni from Syracuse University. Cable radio offered the opportunity to program university sports at a profit. We also had opportunities to promote other events, such as hockey games and eclectic music, that were otherwise not available on mass media. The programming was promoted in sportscasts on the AM station and through direct-mail marketing.

Primitive Cable Mechanics. Our operating costs were minimal. One hundred dollars per month paid for the use of the cable channel. There was a one-time charge of $2,000 for an FM tuner at the cable systems head-end. People placed their radios next to the cable connection on their television sets, or they could actually run a wire from the television to their radio antennas. In those days, we received programming by land line on a barter basis. Today, programming would come from satellite and be microwaved to the cable head-end. Technically, radio has come a long way since the 1980s.

Are Radio's Negative Factors Positives?

The Neighborhoods

Radio is a most flexible niche medium. Radio can target specific neighborhoods—on a daily basis—by simply turning on a remote microphone. As specific locations become more important to advertisers, this remote capability will increase in value. Expect magazines, and some newspapers, to offer individual ads to individual subscribers. Radio can actually be in a neighborhood with a person broadcasting live. This mini-remote can be combined with coupons, order forms, or products distributed to specific people in a designated neighborhood. Imagine the impact, the feedback in a neighborhood, and the results for the advertiser. Radio may become the medium that spawns a new breed of personalities who specialize in neighborhood remotes.

Low Self-Esteem

Since the introduction of television, circa 1945, many radio people have developed a severe case of low self-esteem. Sight and sound were imaged as superior to sound alone. As the years passed, many radio people accepted that perception of radio. According to an ancient Chinese proverb, "A picture is worth a thousand words." This perception is at work in the minds of many radio people, ad agency planners, and clients today. However, there is little reason for low self-esteem when you look at the facts.

The 75% Factor. Perhaps radio is suffering from a mass-market mentality because it is so prolific. As late as 1992, 75% of the world's population received their primary news and entertainment from one source—radio. As governments lose control of the airwaves, as a result of satellite transmission, markets are becoming flooded with the possibility of hundreds of new radio signals. Those stations that become interactive will enjoy an astonishing competitive advantage. Given the ability to interact with inexpensive keypads, radio may become a primary medium in the twenty-first century's niche marketing arena. There are other marketing myths that need to be addressed in order for radio to take a leading place in the next century's media circles.

The Mobility Factor. Radio is the most mobile electronic medium in the world. Radio can broadcast from niches economically. Pocket telephones, wireless phones, and sending a fax from the beach are all possible, using existing radio frequencies. Portable microphones, like portable radios, began to proliferate in the 1940s, and the growth has yet to dwindle. Being unplugged from a socket is as natural for radio as wireless cable may become in television's future. Right now, radio and radio microphones may appear in every nook, cranny, and niche where people find themselves. Radio's mobility—which many take for granted—can be a tremendous advantage as you position radio for the next century.

The Ubiquity Factor. The medium of radio is *everywhere*, cutting across social, political, and economic lines. Radio formats, as you've seen, can target niches like men 25–34. Radio can go to specific neighborhoods and draw niche customers into stores. This ubiquitous medium can interact with the same computer-driven flexibility as any other medium. Plus, radio is available 24 hours a day—no cable necessary, but cable-ready when the right circumstances arise.

Radio's Targetability Factor. Radio has the ability to target specific audiences better than any other media, including cable television. You also have the ability to show, through research, how your station reaches cells such as 25–34-year-old black men who are heavy purchasers of specific brands of beer, like Schlitz Malt Liquor. Cable companies may become even more fragmented as they develop specific niche programming. Most cable companies do not yet have the ability to show their niche audiences in any cohesive manner.

How Small Is Your Niche?

Imagine that your database shows you reach buyers of Mercedes Benz automobiles. You can document the fact using your database. You can show that you have 1,000 listeners who drive Mercedes Benz automobiles. You can show that 50 listeners are in the market for a new $80,000 Benz at any given time. This small group of listeners represents a big reason to advertise. Knowing you reach potential buyers of a specific product is a far better reason to advertise than saying, "We're Number One. We have adults 45–54 who enjoy incomes of $100,000-plus. And, our average quarter-hour rating in this demo is 28,000 from 6 A.M. to 10 A.M." A Benz dealer who can sell to just one of your Mercedes Benz listeners can make a profit. One sale will more than pay for the ad campaign.

Promoting to Your Core Demographics

Herb Hahn, V.P. of sales at Effective Media Services, explains how his company invests a lot of money in television advertising to try to make radio stations more successful. According to Hahn, Effective Media Services buys ad time during specific television programs that target a core audience like adults 25–34. For effective frequency, instead of one ad, they'll buy four or more ads in the same program. This practice amazes many media buyers who reason that a specific program will generate a limited number of gross-rating points. Buying other programs, most buyers feel, would increase the reach of the advertising schedule; they miss the point and the core audience.

Selling Stereos

In niche marketing, our core audience will shrink to very small cells and include similar lifestyles. Effective Media Services is on the right track. The core audience of a major-market radio station could be as small as 1,000 people. These 1,000 people are the heavy, heavy users of the radio station. They may respond to your advertising messages in numbers as few as 100 or less. Selling stereo sets on your air? Selling 100 stereos at $299 each means gross sales of $29,900 for an advertiser. Your advertiser's gross profits could be as high as $14,950, which makes a tax-deductible $2,000 ad schedule look fairly inexpensive. Using the gross-rating–point system is not the most effective method for reaching the heavy, heavy users of $299 stereos. It's up to you, however, to educate media *planners* before you receive calls from media buyers.

Being Different

Radio station operators often clone other successful radio stations. When a format like The Arrow in Los Angeles surfaces, stations across North America rush to copy the idea. Usually they will fail because market conditions in local markets are not the same as they are in L.A. To be a success in niche marketing, you need to be demonstrably different than other radio stations in your market.

General Guidelines

The balance of this book focuses on how you can make your stations different in programming, sales, marketing, and staff composition—and still make a profit. Let's begin by looking at some new research techniques. To help you focus on the new niche marketing realities, the following general guidelines can help you seize profitable niche opportunities.

Niche Marketing Checklist
1. Try to be first in a category in the minds of your listeners and your advertisers—like being the station with the best weathercasts.
2. If you can't be first, don't go head-to-head with competition that owns that position. Try and create a new subcategory such as marine weathercasts.
3. To uncover existing market niches, talk to your prospective core audience listeners individually—not as a group.
4. Find a solution for one listener's needs—for example, marine weather with shrimping reports. If enough people are thrilled about shrimping, you've discovered a potentially profitable niche.
5. If you can fill the newly discovered niche profitably, and no one else is pursuing it, start promoting it.
6. Position your station as the station that "Listens to the Shrimp Run." Create a positive image that people remember FIRST when they think about shrimping (or weather, or marine weather, or any other niche you can find available in your unique market).
7. Become a sleuth. Your advertisers' and listeners' needs are constantly changing. Your sleuthing duties are never over. To succeed in the arena of niche marketing, feed your database daily—or at least weekly. Track your incom-

ing mail. Catalog contest winners. Consider installing a listener comment line. Ask people from which area they're calling.

New Research

Niche marketing covers much more than specialized formats or *narrowcasting* to small groups. You also need to help your advertisers find and promote *their* niches in the marketplace. Successful niching may include using other forms of advertising combined with radio. Some broadcasters fear this will cut into radio's share of the advertising pie. Our research indicates quite the opposite result. There are many exciting things happening in niche research. We'll show you how to take advantage of some of them in the next chapters.

PART TWO
Research

3

Using Niche Marketing Research to Increase Listenership

Great Sums of Time and Money

Traditionally, the media has invested great sums of time and money trying to identify, quantify, and qualify the audience they deliver. Because of more lifestyle choices, you now need to examine your listeners' wants and needs under a microscope. Let's explore some new ways to develop audience information for programming, sales, and marketing applications.

This chapter's objectives are threefold. The first objective is to suggest ways to use the information your station has to further develop your particular programming niche. The second objective is to suggest ways you can gather more information about your core audience. The third objective is to suggest some interactive hardware that will help you develop a dialogue with your core audience. First, let's examine the challenges and opportunities that current research methodology creates.

Using Station Information

The Birth of Average Quarter Hours

In the beginning, there was research. Many ad agencies, retail clients, and radio stations are still dealing with research methods developed in the 1920s and 1930s. Broadcasters, such as Dr. Forest Whan and Archibald Crossley, helped create the terms and concepts we live with in today's media markets. Dr. Whan helped pioneer the concept of Average Quarter-Hours (AQH), which made sense when most radio shows were 15- and 30-minutes long. Crossley's Cooperative Analysis of Broadcasting (CAB) used telephone recall survey techniques. According to Rhody Bosley, partner, Research Director Inc., in Baltimore, quarter-hour measurements have been consistent because they are a compromise between what broadcasters would like versus what advertisers would like. Radio broadcasters would probably prefer time-spent listening data to be reported as average hour, while advertisers would like to have it reported as average minute. The real issue here is the definition of radio listening. The Arbitron Company and the accepted U.S. industry definition of *radio listening* is: "Any radio listening to a single radio station for five minutes or more in a given 15-minute period." The British definition is: "Any listening in a quarter-hour segment—but, reported in Average *Half*-Hour." Note how much better U.S. radio is with its definition because three stations can get credit for the same quarter-hour.

Group Mythology

To succeed in the business of niche marketing research, however, it will help if one starts by trying to shed tendencies to place people in large, generic groups by age and sex, such as women 25–44. Take the Hispanic group, which really isn't a group; it is made up of Cuban, Mexican, Puerto Rican, Guatemalan, Castellan, Basque, Mallorcan, Argentinean, Paraguayan, Venezuelan, Bolivian, Panamanian, Chilean, Ecuadorian, and Columbian lower-, middle-, and upper-class people. These diverse types don't fit neatly into one ethnic label like Hispanic or Spanish-speaking persons; however, some researchers and marketers still want to lump Hispanics together as a group. Each nationality represents *at least* one potential niche. According to a July 1994, *U.S. News and World Report* article, "Hispanic Radio Today," "Many polls show Latino attitudes and values are not very different from those of other Americans." Hispanics do, however, retain their own cultural preferences in music, food, clothing, art, literature, and radio (Figure 3.1). Your job is to dig beneath the surface to see if those cultural differences can become profitable niches.

Up-to-the-Minute Research

In the era of niche marketing, the quarter-hour compromise may no longer be the focus. Niche marketing will require up-to-the-minute research instead. How recently the research was conducted and how much the market may have changed since then will become more important to advertisers and radio stations.

Consumption Behavior

According to Bosley:

> Radio listening behavior is a personal preference. In the past, researchers believed that various groups of individuals selected different stations based on their age, sex, location, and ethnicity. Many researchers believe that age and sex are the prime determiners of purchases and social behavior. But others are quick to point out that a woman who has a baby at 21 and a woman who has a baby at 45 both have the same consumption needs for the baby.

```
                    Spanish 43.1%
```

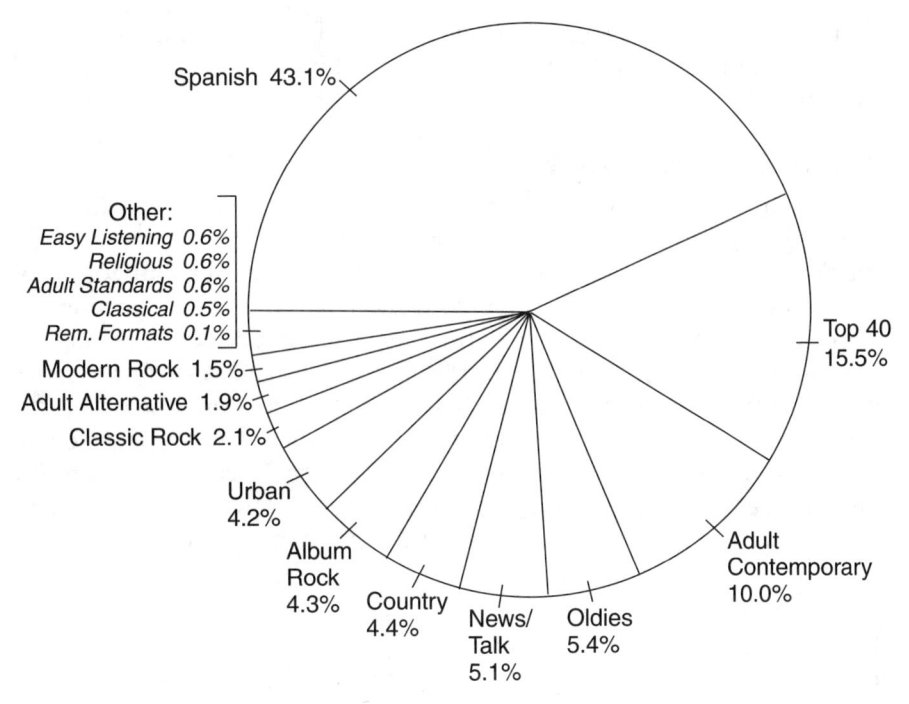

FIGURE 3.1 Hispanics' radio listening habits in 1993.
Source: "Beyond the Ratings (BTR)," courtesy of The Arbitron Company.

Pie chart segments:
- Spanish 43.1%
- Top 40 15.5%
- Adult Contemporary 10.0%
- Oldies 5.4%
- News/Talk 5.1%
- Country 4.4%
- Album Rock 4.3%
- Urban 4.2%
- Classic Rock 2.1%
- Adult Alternative 1.9%
- Modern Rock 1.5%
- Other: Easy Listening 0.6%, Religious 0.6%, Adult Standards 0.6%, Classical 0.5%, Rem. Formats 0.1%

The babies may have the same basic consumption needs but what the 21-year-old and the 45-year-old can provide for their babies may be very different. Radio consumption behavior and product consumption behavior both need to be better understood.

The Difference Between 30-Year-Olds

In the demographic-cell mentality, marketers assign group numbers, like women 25–44, in media plans that target supermarket shoppers. Bob Jordan, president of International Demographics, Inc., asks an incisive question, "What does a single, 30-year-old woman have in common with a divorced, 30-year-old woman with three children?" From a supermarket's niche marketing perspective, the answer may be: "Not much!"

Database Mountains

As databases grow, radio people are being buried under reams of information. Unfortunately, most don't know what to do with this mountain of intelligence. Many broadcasters believe they haven't the time or the facilities to process all the information they've compiled about their listeners.

There are software packages, however, that will perform the services for sales, programming, and marketing applications to help process this gold mine of data.

Using In-House Data in New Ways

What is an in-house database? We're glad you asked. The sources for your in-house programming database are listeners who win on-air contests, listeners who write letters to your radio station to comment about your programming, and listeners who call your radio station to ask questions or make comments about your programming. You can use databases in new ways to help learn more about your listeners. You'll advance faster if you avoid many of the old practices that surround current research techniques. For instance, you can eliminate the practice of focusing on cume information to emphasize a mass audience. *Cume,* or circulation figures, are holdovers from the heyday of newspaper research. Advertisers don't buy the cume of a particular daypart or the cume of a week.

The 100% Error Factor

Audience surveys offer fairly reliable information in broad demographics, such as persons 7 to 12-plus. Niche research, however, requires a look at much smaller cells, such as men 20–24. In mass-market research you'll see disclaimers like "There may be a 100% error factor in a particular demographic cell." For example, an 18–49 cell—which consists of people 18–24, 25–34, 35–44, and 45–49—may have a 92% confidence factor. However, a smaller cell, men 18–24, could be 100% inaccurate because only a few males in that age range actually participated in the research. In niche marketing, you'll be dealing with very small cells.

When we consulted Gerry Boehme, vice president and Director of Radio Research at Katz Radio Group in New York City, he agreed that in some extreme cases the projected estimates—so common in mass-market research—may be based on as few as three people in a niche demographic cell such as women 18–24. In fact, three people can represent 100% of the entire 18–24 market sample in areas with populations of 100,000 or less! To further confuse the research, it is common practice to *weight* undersampled demographic cells (e.g., three people may carry the weight of a sample of 30). In this chapter, you'll see how to gain a more reliable picture of your target audience.

Research Professionals

Professional researchers may discover important data but, after the survey is completed, your market keeps changing—constantly. Certainly, it's prudent to suggest full-time researchers for specific needs. But don't feel you must always use them. Even though we own a marketing and research company, we advocate an ongoing program of your own in-house research.

Research companies often use unskilled workers for phone and face-to-face interviews in malls. Then, after the interviews are completed, workers tabulate and analyze the information. Conclusions must be drawn from the research. The information passes through several hands, and numerous steps may be performed by different people. By the time the report is completed, it has often traveled through the hands of many workers—the margin for error is enormous.

Consult the Ultimate Source

Your listeners are the ultimate source—consult them. Engage in listener research continuously. Start with information you already have. Most stations have a list of recent contest winners who had to listen to the station to win. Because it's less expensive and time-consuming to keep a listener than to gain a new listener, this is a good place to start. There are many ways to approach these valuable radio station consumers. Let's look at two of the easiest.

One Listener at a Time. If you're serving or looking for niches, it is important to talk with listeners individually—not as a homogenous group. Seasoned management consultants would rather talk to two consumers for two hours than review the results of a 2,000-person survey. Connecting with your radio listeners can yield far more useful information than blindly contacting thousands of people.

Use the Telephone. You can survey your contest winners by calling them anonymously. You'll find that there's less bias in the responses if your station is not identified. Some broadcasters feel that listeners will be flattered if they know the radio station values their comments enough to call and interview them. While your listeners may be flattered, as human beings who are taught from the cradle to be polite; there's a much greater chance that you'll be given answers intended to please you instead of receiving information that will help you. You'll be using many of the same procedures used in a conventional telephone coincidental. Prepare a 2- or 3-sentence script to explain the purpose of the call. Have 5 or 6 questions that can be answered with 1- or 2-word answers. If you have an all-news format, or a heavy news commitment, you might ask some of the questions on the sample form in Exhibit 3.1.

Some Limitations. Although the telephone survey is a convenient way to glean information from your listeners, the data will be limited because the interview will be brief. Also, you're limited because you can ask mainly questions that can be answered by a Yes or No, or a 1- or 2-word

Exhibit 3.1

Sample Telephone Survey

Phone: _____ Date: _____
Listener name: _____
Address: _____
City: _____ Zip code: _____
Block: _____

Hello. My name is Jody Watts. I'm with Seaway Research. I'm not selling anything. I just need to ask you a few questions about radio news programs. Will you help me? ☐ Yes ☐ No
   (If no, go to number 9.)

1. Do you listen to the news on the radio? ☐ Yes ☐ No
     If no, go to number 9.
2. Are there certain days that you listen to the news? ☐ Yes ☐ No
     If yes: Which days do you listen?
     ☐ Mon ☐ Tue ☐ Wed ☐ Thu ☐ Fri ☐ Sat ☐ Sun
3. What times do you listen to the news on the radio? _____
   _____
4. Which radio station's news do you like best? _____
   _____
5. Who's your favorite radio newscaster? _____
6. Do you think radio news stories are too long? _____
   too short? _____ just right? _____
7. What is your *primary* source of news? _____
   _____
8. In which group do you fit?
     ☐ 18–24 ☐ 25–34 ☐ 35–44 ☐ 45–54 ☐ 55–64 ☐ 65–74
     ☐ 75–84 ☐ 85+
     ☐ male ☐ female
9. Thanks so much for your help. [*Hang up.*]

---

reply. The last question, however, could be open-ended. Allow respondents to say anything they wish and record their comments verbatim.

Use the Mail.   You may choose to send a direct-mail piece to the listeners in your database. Many companies conduct niche marketing surveys using the mail. If you've personally purchased a new product lately, within a week you probably received a questionnaire in the mail asking your opinion of the product. Often, you'll be entered into a contest to win free products if you complete the survey. Stations can develop a questionnaire for their listeners. Offer to enter their names into a contest for a prize if they complete your form. Some companies offer no prize, but the response rate will improve with an incentive.

For Best Results.   You'll need a very specific objective to achieve the best results. In the sample in Exhibit 3.2, the objective is to discover more about the musical tastes of your core audience. Rhody Bosley feels that music directors at client stations should perform ongoing listener research. Bosley suggested some of the questions for this listener survey form.

Exhibit 3.2

Tell Us What You Think

Your comments can help us in our continuing effort to improve our radio station for your enjoyment. Please take a few minutes to complete this brief form and drop it in the mail. In return for your time and effort, your reply will be entered into the Grand Prize Drawing for a dinner for two at *The Steakhouse*.

1. Who are your three favorite musical artists?_____
   _____
2. What are your three favorite songs?_____
   _____
3. Do you have a favorite radio personality?_____
4. Who is it?_____
5. In which group do you fit?
   ☐ 18–24   ☐ 25–34   ☐ 35–44   ☐ 45–54   ☐ 55–64
   ☐ 65–74   ☐ 75–84   ☐ 85+
   ☐ male   ☐ female
6. When do you listen to our radio station?
   ☐ early morning   ☐ midday   ☐ late afternoon   ☐ night
7. Usually, where are you when you listen to our radio station?
   ☐ in my car   ☐ at work   ☐ at home   ☐ some other place
8. What is your favorite radio station?_____
9. What is your second favorite radio station?_____
10. What is your third favorite radio station, if you have one?_____
11. Which station do you spend the most time listening to?_____

Thanks so much for your help. Good luck!

Tabulate Your Responses

Skip Finley, president and general manager of WKYS-FM, Albimar Communications in Washington, DC, has uncovered significant data about his upper-income urban audience. Skip has mastered the mechanics of block-group coding. Here's how Skip defines it: *block-group coding* is groupings of physical city blocks; they are Census-defined. You may find as many as 250–500 households in a block grouping.

Birds of a Feather Flock Together. Block groups are usually made up of very homogenous characteristics. For example, you may find a block group of people who have a median income of $45,000, 73% of whom have a high school education, 22% of whom are homeowners, 34% of whom are Chinese Americans, and 65% of whom are Caucasians. You'll find a lot of census, audience, lifestyle, and product information, which is usually available at the zipcode level. A typical zip code, however, may be made up of multiple block groups.

The Difference Between Block Groups. Block Group One may include people whose median income is $175,000, who are almost exclusively Caucasian, and are high school–educated, with 47% of them owning homes. Block Group Two's median income may be $25,000. Block Group Three may have a median income of $62,000 and Block Group Four's median income may be $40,000. Block groupings are important to niche marketing because they allow radio stations to focus more closely on the segment of the market that is most likely to listen to a particular format. Once your *hot* block groups have been identified, you can study those groups under a microscope. In a never-ending quest to understand your audience and what motivates them to listen, block groupings are a very helpful tool.

Page M6. On page M6 of every Arbitron book, you'll find the following description of how information is gathered to create block-group coding. Arbitron uses a market segmentation system from Claritas, Inc., called PRIZM™ (a registered trademark of Claritas). This system is designed to help marketers target consumers and profile markets and audiences by lifestyle. Claritas analyzes key demographic characteristics from the U.S. Census and hundreds of millions of actual consumer purchase records to classify each neighborhood in the United States into one of 62 distinct PRIZM clusters.

Among the characteristics analyzed are income, value and type of housing, marital status, presence and ages of children in a household, ethnicity, urban-suburban-town-rural location, age, sex, occupation, level of education, as well as new car registrations, magazine subscriptions, real estate transactions, and financial data. Claritas updates PRIZM annually. The 62 unique PRIZM clusters are organized into 15 broader social groups. Each group is identified by a group code, which denotes a basic neighborhood type. When linked to market and radio measurement data, this geodemographic model produces descriptive audience information.

Each PRIZM group's composition in this metro for persons 12+ is compared to the group's national composition. The index of concentration compares the metro's composition with the national composition for each PRIZM group. An index of 100 indicates that the market composition is twice that of the nation. For more information about PRIZM, call (703) 812-2700.

Focusing on Your Core

As a programmer, you'll probably want to conduct more specific research about your core audience. The information a service like Arbitron supplies is often helpful but doesn't give enough of the one-to-one research needed to be a successful niche programmer. During a survey, diary placement may yield only a few responses from your hot blocks. You'll need more information about your core listener. Arbitron can't be expected to survey each local market or designated market area (DMA) in the country. You'll need to conduct your own customized, local research.

Merging. Arbitron's block-group coding should prove useful at the metro level. You can merge your listener database with Arbitron's block-group coding on maps of your market. Then, when you've finished tabulating the results of your in-house surveys, categorize respondents by block groups. Today, several hundred households comprise a block group. In the future, however, it may represent as few as 50 people. Now that the results are tabulated, you can pinpoint people in a certain area who think that news programs are too long, as well as people in another group who feel that radio newscasts are too short. Next, you'll need to decide which niche is more advantageous to your overall programming goals and your core audience.

For Example. After surveying for niches, you may find that you have an accumulation of various niches. You'll need to appeal to similar niches without causing them to tune out. You'll want to understand your core audience intimately. If the core target is affluent 45–49-year-old females who listen to news, you can make intelligent programming adjustments. This data also provides an information bonanza for the radio station's sales department; we'll cover this in the next chapter.

Information Retrieval

Many people perceive the Information Superhighway, or I-Way, as interactive, but it isn't yet. We may be well into the twenty-first century before real-time interaction is possible. Some banks offer what looks like an interactive system, which allows customers to check balances, deposits, and receive a current report on the last five checks that cleared. These are basically information-retrieval systems. Some systems may allow a customer to electronically select an item to purchase, indicate an opinion from a preselected list, or vote for their favorite radio personality. The technology, however, is still a process of information retrieval. To achieve a microscopic picture of your niche market, we advocate as much one-on-one interaction with your listeners as possible.

Interactive Computer Limitations

Our current interactive computer systems offer some exciting possibilities; however, two-way dialogue is still limited. Even computer bulletinboards, which allow two-way communication, experience real-time delays. Radio stations will still need program directors and research people who can meet with listeners one-on-one well into the twenty-first century.

The Difference Is Listening

One of the best methods for true interaction is the focus group. We'll discuss focus groups in detail in Chapter 6. In niche marketing, your ultimate objective will be to monitor each core listener and find what delights—not just pleases—them.

4

Using Niche Research to Increase Sales

Sleuthing

In the late 1940s, Ray Herweg, a Chicago advertising agency executive, decided to solicit the Sanford Ink account; Sanford was a manufacturer of inks, mucilage, and stamp pads. His research started at the two largest fountain pen–repair shops in Chicago. Herweg's question was, "What could be improved in the manufacture of inks to increase customer satisfaction?" The pen-repair mechanics told Ray that ink tends to gel after several months in the bottle—resulting in clogged fountain pens. Customers were unhappy with the performance of their pens, but didn't really understand the problem that created their complaints. It was really very simple: The most popular ink bottle sizes in the 1940s were 2- and 4-ounce bottles. Those bottles took about six months to use—almost guaranteeing to clog the pens as the bottle was depleted.

The Traditional Advertising Approach

Several major ad agencies in Chicago were also interested in the Sanford account. In those days focus group research was gaining in popularity. The groups seemed to offer an efficient way to test advertising and product appeals. The competitive ad agencies created campaigns that focused on the "quality" and "value" of the Sanford product line. These approaches may remind you of the quality-and-value bromides we hear in countless radio and television ads today. The ads that tested best were presented to Sanford by Herweg's competitors. In this instance, focus group research findings mirrored the predetermined conclusions of the advertising agency's executives. In the 1940s, many marketing people believed that consumers mainly wanted to hear about the quality and value of a product. So, the ad campaigns centered around quality and value, and they tested well.

Unorthodox Research

Herweg's approach was different; he built his campaign around the unorthodox sleuthing he did at the pen repair stores. He proposed that Sanford reduce the size of the ink bottles from 2- and 4-ounce to ¾-ounce bottles. This meant a complete retooling of the assembly line. Imagine, making a presentation to one of your prospective clients suggesting that they retool their product line and completely restock their store! The Sanford Ink people listened—and acted. The new bottles were called "Ink Wells." A line of ten colors marketed in dimpled ¾-ounce bottles was suggested. Most ink manufacturers sold a basic line of blue and black inks in that era. This was revolutionary. The ad copy platform was built around the fact that Sanford Inks helped prevent pens from clogging. The theme was, "All the ink you can keep fresh."

Initially, this bold approach met with opposition from the Sanford sales force. Ray offered to make calls on leading clients with the salespeople, however. This is the kind of face-call marketing that radio sales and management people perform today, but it was rare then. The logic of the campaign and the added marketing efforts landed the account for Herweg. In fact, Ray kept the account until he retired over 20 years later.

Are we suggesting that niche research is nontraditional? Yes. Can this nontraditional, local-research approach help stations keep accounts longer? Yes. Is it possible to sell long-term contracts using niche research? Yes, and it's easier.

Broadbased Demographics

Retailers, ad agencies, and radio stations continue to use broadbased, mass-market demographic cells to define their advertising targets. Let's go one step further and narrow the demographic base, using supermarkets as an example. Ask a marketing director for supermarkets about the demographic of their target consumer and you'll most often hear "females, 25–54." Examine an even narrower age range of supermarket shoppers such as 18–34-year-old males. Is a single, 18-year-old male buying the same products as a married 34-year-old male with a family? Probably not. Today, you need more targeted campaigns and more in-depth research.

Getting Started

To start on a niche marketing path, find out what appeals to your advertisers' customers—one customer at a time. For example, let's look at one possible niche for a shoe-store client. Perhaps your in-the-field research has shown you that one teen customer has been interviewed. She's told you that she owns ten pairs of shoes; nine pairs are the same color—black.

Eight of the pairs are flats. Will these statistics match answers from other teen customers? Now, determine how many other teens would be delighted—not just satisfied—if given choices in your client's shoe store that correspond with your research. If that number is large enough, you have a potential market niche. This is a far better approach than breaking down mass markets into smaller segments.

Breaking down mass markets implies that you have products and services in search of markets. For example, the shoe store advertises that it is having a storewide sale on shoes for men, women, kids, teens, and senior citizens (Figure 4.1). The store owner is making one offer to multiple segments of the population. The shoe-store advertiser assumes that the varied customer base will be most motivated by price. For niche market advertising campaigns, however, you need to find out what *thrills* each customer—not what might be *generally* pleasing to them. Each group of buyers—whether it's teens, kids, men, or whomever—has different reasons for buying shoes at that store. For instance, a majority of women go into that shoe store because it has a reputation for carrying a wide selection of high-heeled dress shoes. This would not be the same reason that senior citizens, teens, or any of the other three segments goes into Paul's Shoe Store. Five different ad campaigns—to attract each of the segments—may make the most marketing sense.

The Average Customer Myth

Many advertisers believe that they have an *average customer*. What we have described as an average customer until now may really be five or six different customer types. If you look at an *average customer* under a microscope,

Paul's Shoes
The Family Shoe Store

Clearance

50% Off ALL Shoes!

MENS   WOMENS   TEENS
SENIORS   CHILDREN

555 Main Street   Open Mon-Fri 9-5
Big City, Big State   Sat   9-5
555-4321   Sun   closed

FIGURE 4.1   Mass-marketing approach

you might see that what you have thought of as an average customer is really composed of a lot of different kinds of customers with very different needs and different buying motives. To understand your clients' customers better, take the word *average* away from the word *customer*. There is no *average customer* just like there is no longer a mass market for goods, services, and radio stations. In the niche marketing future, you'll need to take a much closer look at your advertisers' customers (Figure 4.2).

Ethnic Diversity. As markets continue to fragment and diversify, ethnic composition becomes more important. Consider the situation in Singapore—76% of the population is Chinese, 15% is Malaysian, 7% is Indian, 2% is "other." The language of commerce is English. The major radio format, so far, is an English-language format, which has been influenced by the Chinese majority. In 1994, the Malaysian government's broadcast regulations diminished and the number of niche stations grew. The Singapore example suggests that an "Asian" may be a number of very different types of people. Asians, like African Americans, Caucasians, and Hispanics can no longer be packaged into large ethnic blocks. If you want to create more effective advertising copy and media plans, you need more focused research.

Ethnic Weighting. Meanwhile, back in North America, you're still deluged with mass-market research techniques. These techniques include ethnic weighting for large blocks of people such as Asians. Ethnic weighting may further distort the conclusions of niche demographic cells. For example, *one* person—in sometimes undersampled ethnic groups—may represent 100% of an 18–24 cell. What follows is an example of how you can break out of current research habits and still help clients grow using more innovative research.

Fast Forward to the Present

Bob Parker, president/COO of Sanford, is intensely interested in niches. He doesn't use mass marketing for his company's product lines. According to Parker, at one time Sanford's Sharpie® Marker—which owns a majority of the marker market—was advertised in mass media. Now, Parker feels that direct calls on the office products buyers for companies such as IBM and AT&T is the most efficient way to reach the consumer of markers. So far, Parker's logic is impeccable.

Radio's Multimedia Challenge

Radio now has a multimedia task. The job is to show how direct calls, direct mail, and daily local radio advertising can increase sales for San-

FIGURE 4.2 The "average" customer

ford. The opportunity in locations where large companies buy markers and pens can be enormous. In the case of companies like Sanford, you can call on the office products buyer in a business and learn what interests them about Sanford products. You can ask what new products they are looking for or using. Given the right circumstances, people love to share their expertise and give advice; *and*, advice is what you're after.

Companies that understand the importance of marketing niches are eager for more information about their heavy users and potential heavy users. Remember the frequent-flyer airline niche example discussed in Chapter 1? Airlines have found that approximately 5% of the adult population makes approximately 70% of all air trips. Delta Air Lines is eager to learn more about their heavy users. Your advertisers are also eager to learn more about *their* heavy users:

- Ask the office products buyer how the company consumes a product or service.
- Ask which seasons—if any—are the best times for using the products in question.
- Ask what your potential client can do to improve its products or services.

- Ask why they buy the product or service and what would make them buy a competitor's goods or services.
- Ask about packaging, delivery times, and pricing.

You'll get insightful answers and a new perspective that an official company representative won't receive. You can include this information in a presentation to your prospective client. Given the quality of the answers, you have become part of the marketing team. You'll help save the product manufacturer many thousands of dollars in direct-call and direct-mail expenses. Are we suggesting that you'll replace direct calls or direct mail? No. You will, however, increase the value of your radio product. You'll sell more time to clients because you've made your radio station a valuable part of their research and marketing effort.

A Different Kind of Marketing Presentation

Now, make a presentation to the company in question. You know that radio reaches office managers with great frequency. Radio can also reach the people who influence the purchase of office products. How many blue-chip companies are in your market? How many large companies in your market consume volumes of office products or other goods and services? Radio reaches the ultimate consumers—right in their offices. Your customer research can include new ways for office workers to use—in this example—markers. The Sanford salesperson is on-site for a short time and direct mail is effective for just a few moments, but radio can be in the office every day for hours. The question you need to ask is, "Will this research-and-copy approach be a profitable niche for my radio station?"

Secondary markets are created by retailers like Office Depot. Here's another avenue to use niche research to demonstrate the power of niche formats combined with niche marketing. In the case of markers and pens, this means interviewing the end-consumer in or out of stores—yourself. (Sample research follows.) This means talking with the office products buyers in large companies in individual markets. This means presenting our findings to people like Bob Parker at Sanford. Sanford is a consumer-driven company looking for the results of new sleuths like yourself. Niche marketing research is different. The traditional research approach has been too general for the new era of niche marketing.

Share-of-Advertiser

In the new era of niche marketing, you'll form an alliance with each of the advertisers with which you work. They will begin to study the information you bring to them. Instead of focusing solely on the level of your billing this month, without regard to whom that advertising was sold, niche marketing sales managers will gauge your success by the *increase* in an advertiser's consistent expenditures with your station. Your success quotient will be determined by increasing share-of-advertiser, not share-of-market. The most important part of your relationship with each of your clients in the niche marketing future will be asking questions and getting feedback. What do advertisers really want? What does this advertiser want?

When you begin to think about increasing each of your advertiser's spending with you, new ideas will occur to you. You'll see all sorts of ways to upgrade your clients. Instead of being bewildered by these new research methods, you'll want more and better questions to ask. You'll work with fewer advertisers but each one *you choose to work with* will spend more of their budget with you. You'll get to know each of your clients and their business very well. You will sell more advertising—without reducing rates—while *pleasing* your advertisers. Follow the system and sell more to your best clients—which will make them value you more. The main ingredient in this formula for success is to know your advertisers extremely well. The first step is to decide which advertisers will never buy from you so you can stop spending emotional and physical effort on them. Invest your efforts in your loyal advertisers so you can work toward increasing their expenditures with your radio station—now.

Step One

As a radio sales representative, your biggest growth area is local direct retail business. Let's examine the sequence for niching for a prospective retail advertiser. *Before* you try to sell anything to an advertiser, interview your prospect using the Niche Marketing Analysis: Part I (NMA: I) form in Exhibit 4.1 or your own modification of it. The analysis interview is the first step in starting a new relationship with a prospective client or upgrading a relationship with an existing client. This interview system will open doors for you because you aren't selling anything—you're just asking for the opportunity to ask questions. The advertiser's answers to your questions will help you make more focused and effective presentations.

A Confidence Builder

As you perform niche marketing analyses, you'll gain more confidence in your ability to make marketing suggestions that will enhance your clients' business. Use the interior design motto of "form follows function." A designer decides how to decorate a room based on what people do in that room. Asking certain questions of the room gives the designer the path to follow. The answers to the NMA: I questions will tell you which marketing direction you need to pursue. For example, an independent car care center owner felt he didn't need radio advertising. After all, his yellow pages ad listed all of his services; anyone who wanted service could look in the

Exhibit 4.1

Niche Marketing Analysis: Part I

Company: _____ Date: _____
Address: _____ Phone: _____
E-mail address: _____
Other locations: _____
Decision maker: _____ Title: _____
   Time with company: _____ Year company established: _____

CUSTOMER PROFILE

Of your customers, there's a group that we call your heavy users. They may visit your business more often and they definitely buy more of your products. What is their age range? ☐ 12–17 ☐ 18–24 ☐ 25–34 ☐ 35–44
  ☐ 45–54 ☐ 55–64 ☐ 65–74 ☐ 75–84 ☐ 85+
  ☐ **Females** make buying decision ☐ **Males** make buying decision
  ☐ **Couples** make buying decision

Describe the lifestyle(s) of your customers: _____
  For example, are they:
  ☐ especially active ☐ families ☐ homeowners ☐ apt dwellers
  ☐ boaters ☐ sports participants ☐ movie-goers ☐ dieters
  ☐ white collar ☐ blue collar ☐ pink collar ☐ agri-business ☐ military
  Other (specify): _____

How far do your customers travel to get to your store? _____
How much do they spend on an average visit? _____
Are the majority of your products/services impulse buys? _____
What is the average length of time your customers are "in the market" for your products? (E.g.: What is the time span between when customers decide they need your product and when they actually make the purchase?) _____
Why do customers shop here? _____

POSITIONING ANALYSIS

When people think of your company's name, what image do you want to come to mind? _____

What is the number one misconception customers have about your company?
_____

What are the most important things people should know about your business?
_____

Who is your largest competitor? _____
  Why? _____

What are the most important benefits of shopping at your business? _____

What makes you different from your competitors? _____

PREVIOUS ADVERTISING PROFILE
Newspaper(s)

_____

_____
(list papers)

Why is newspaper advertising effective for your business? _____

Are there disadvantages? _____

Who is reading your ad in the newspaper? _____

What percentage of your total advertising budget do you spend on ads in newspapers? _____

Yellow Pages

_____

_____
(list phone books)

Why is yellow pages advertising effective for your business? _____

Are there disadvantages? _____

Who is reading your ad in the yellow pages? _____

What percentage of your total advertising budget do you spend on yellow pages advertising? _____

Radio

_____

_____
(list radio stations)

Why is radio advertising effective for your business? _____
Are there disadvantages? _____

Who is hearing your ad on the radio? _____
What percentage of your total advertising budget do you spend on radio? _____

Exhibit 4.1 *continued*

Cable/Over-the-Air Television

_____

_____
(list television stations)
Why is television advertising effective for your business?_____

_____

Are there disadvantages? _____

_____

Who is seeing your ad on television?_____
What percentage of your total advertising budget do you spend with television stations?_____

Direct Mail

_____

_____
(list when direct mail sent out)
Why is direct mail advertising effective for your business?_____

_____

Are there disadvantages?_____
Who are you reaching using direct mail for advertising?_____
How do you generate your mailing list?_____
How often do you update your list?_____
Do you purchase your mailing lists locally?_____
How many mailings do you distribute each year?_____
   each quarter?_____ each month?_____

Other

_____

_____
(e.g., billboards, inserts, magazines)
Why is _____ advertising effective for your business?_____

_____

Are there disadvantages?_____

_____

Who is seeing your ad on (e.g., billboards)?_____

ADVERTISING GOALS

What do you want your advertising to accomplish?_____

_____

Do you base your advertising budget on a percentage of last year's sales or on projected sales for next year? _____

Approximately how much do you spend each year on advertising? $ _____ or
☐ 1% of sales   ☐ 2%   ☐ 3%   ☐ 4%   ☐ 5%   ☐ Other: _____ %

When do you make advertising decisions? _____
   Exp: ☐ When the need arises   ☐ weekly   ☐ monthly
         ☐ quarterly   ☐ yearly

Who sets the advertising goals in your company? _____
_____

How many people make the final advertising decisions? _____
   Who are they? _____
_____

PROMOTIONAL SCHEDULE

When do you advertise?   ☐ near holidays   ☐ preseason   ☐ clearances
   ☐ your company's anniversary   ☐ other (specify): _____

What is the most successful promotion you've ever had? _____
_____

   Why was it successful? _____
_____

SALES CYCLE

Every business has stronger months and weaker months. What is your strongest month?   Weakest month?
   Which are good months?   Fair months?

(**S** = Strong, **G** = Good, **F** = Fair, **W** = Weak)

Jan _____   Feb _____   Mar _____   Apr _____   May _____   Jun _____
Jul _____   Aug _____   Sep _____   Oct _____   Nov _____   Dec _____

CLOSING

Do you have any copies of trade magazines I may borrow? _____

What would you like from media reps that you're not getting now? _____
_____

Are there any other questions I should have asked? _____
_____

I need approximately seven days to conduct my in-house research and prepare a presentation. Is this time next week good for you?   ☐ Yes   ☐ No

Next Appointment   Date: _____   Time: _____

phone book and see his half-page ad. Tina, the sales rep, listened patiently and then explained, "This interview will help me understand your niche in the market so I can study it, research it, and come back with marketing suggestions." He may have been curious, however, about a sales rep who just wanted to ask questions about his business instead of pitching her special-offer-of-the-week. He agreed to the interview, and Tina asked the NMA: I questions.

Successful Interviewing

The most successful interviews are the ones where you learn the most. You'll receive the most information when your advertiser is relaxed and has confidence in you. Anyone who has ever been interviewed has wondered if the interviewer might misinterpret his or her words. You may want to begin your session by assuring the advertiser that he knows all the answers to your questions. You, however, don't know a lot about his company and his customers, but you would like to learn about them. It is the interviewer's responsibility to help the interviewee give specific information and elaborate on answers and to put the responses in context with the larger picture of the company's overall business.

Asking Questions

When you look at the first set of questions on the NMA: I form, you may already know the answers to some of the questions because you know the name of the company and its address. Overcome the temptation to fill in the top of the form before your interview. Asking these nonthreatening questions is a good way to begin the interview. The authors have performed several thousand NMA interviews and it always amazes us when we learn new information from a prospect—that we *thought* we knew in advance. You will also want to make sure you have the accurate title of the person you're interviewing because you'll want to use his title on the cover sheet of the presentation on your return visit. This interview form is designed to begin with nonthreatening questions. Questions that may seem threatening in any way appear later—after you've had an opportunity to establish your sincere interest in the advertiser's needs. If you detect that the interviewee feels threatened by a question you've asked, explain why you need the information. Most people will relax and give you the information.

Eliminate Generic Words

Generic terms can hinder communication. For example, when you ask an advertiser the age range of his customers, you may hear a response like "My customers are young." At this point, you need to ask, "What does *young* mean to you?" (To help your advertiser relax you may make a comment such as "I find that the term *young* changes as I get older.") If your prospect tells you that he makes his advertising buys in the *spring* and *fall*, you will need specific months. You may ask "What months would that be?" The spring season can vary by several months depending on the products or services being offered. When asked why people shop at an advertiser's store, too many times the response has been "I have *quality* products." In order to help this business manager, ask him to give you an example of what *he* means by quality. You may hear excellent reasons for shopping at this store that you can use in his radio ads. Remember, you may be on a fact-finding quest but you'll both enjoy the experience more if you approach this interview as two friends trying to get to know each other better. Your interviewee will be getting to know you as a result of how you handle yourself during the discussion.

Customer Profile

Explain to your prospect that you're going to focus on his customers for a moment. Experience has shown that when you ask about the age range of a prospect's customers, answers sometimes range from 12–84. Advertisers are often proud of the vast range of their customer base. If you've been in sales very long, you know that a customer base of 12–84-year-olds is not an accurate picture of the prospect's heavy users. It is your responsibility to probe for more definitive information. Paraphrase your question; for example, you may say "That's terrific that anyone from 12 to 84 can use your service but I need to take a closer look at those customers I call your bread-and-butter customers—the ones that come in here more often and buy more of your products (or services)." At this point, you'll find that most people will narrow the demographic of their customers, giving you a more realistic description of their heavy users. And, even if you *think* you know the answer, you still need to ask "Are most of your customers female or male? Do couples make the buying decision? Is there a similarity in the lifestyles of your customers?"

How Far. You need to know how far customers travel to get to the advertiser's store. When you make your presentation, you may choose to compare geographically where your listeners are located and where the interviewee's retail trading zone is located. Your comparison map doesn't have to be elaborate. Computer graphics are nice but not necessary. If you have a listener-response map at your station—not a signal coverage map but a map that shows where your station has had response from listeners—you may simply draw a circle in color around each store location. This is a simple but powerful way to show the advertiser how your station's listeners geographically reach their bread-and-butter customers. Knowing how far a customer travels to get to the prospect's store will also help you when it's time to write a client's commercials.

Why Shop Here? "Why do your customers shop here?" This is an extremely important question. In order to get *other* customers to respond to advertising, you must find out what motivates the company's present customers. An advertiser may not know exactly what motivates customers to shop at his store. When that happens, probe for more information by discussing the needs customers are trying to fill by purchasing the store's products (or service).

Positioning Analysis. Where has the prospect positioned his company in the minds of consumers? Is the position he owns in the marketplace the one the company wants to own? You'll quickly find whether the advertiser has made a commitment to building a specific image. If you believe that the business has a different image from the one the decision maker is describing, wait until you make your presentation to discuss this topic. At that meeting, you will have had time to think about a logical and tactful way to show your prospect another perspective.

Misconceptions. "What is the number one misconception people have about your company?" Every prospect you query knows the answer to that question. Decision makers spend a lot of time thinking about their business. They analyze their customers' perceptions and misconceptions. Frequently, advertisers hear their customers discussing beliefs they have about their business. Store employees also field responses from customers that uncover misconceptions about the company. The need to change misconceptions is an important reason to advertise.

Previous Advertising Profile

Many times sales reps have relationships with advertisers for years and never get around to discussing how the advertiser uses other media. It's *OK* to discuss your prospects' and clients' buying habits with other media. If you are committed to being the best niche marketing consultant possible, you'll want to gain as much knowledge as you can about all media. You won't be gaining this knowledge in order to criticize all other media; your advertisers can make buying decisions for themselves. You'll use your knowledge of multimedia to help clients make more effective buying decisions. Allow your advertisers to talk about how they use other media. How can you truly become their marketing partner if you can't see the whole landscape of their marketing efforts? Integrated marketing is the most successful.

Reassuring the Prospect. The NMA: I form was developed and refined over many years as a tool to help salespeople and advertisers improve communications and help you make a more meaningful and logical marketing presentation at your next meeting. If someone questions why you need to ask about other media, you can explain. Tell the prospect (or client) that you need to take a look at *all* of the company's advertising efforts because you will go back to your station and study the previous advertising profile through a niche marketing microscope. When you return with a presentation, you'll suggest ways to fine-tune and/or integrate the client's marketing efforts to more effectively reach his customer niche.

Newspapers. In the newspaper section, ask "Who is reading your ad in the newspaper?" Often, you'll hear "I'm not sure." At the other end of the spectrum, you may be told "Everybody." Either answer will not give you enough information to help this person. You may need to ask specifically if there is an age group that the advertiser believes is reading the newspaper ads. Regardless of the answer to this question, when you hear a lot of specific information from a retailer, you will very likely be hearing what newspaper reps are telling him—in order to sell *their* product. If the advertiser believes glaring misconceptions, this is your opportunity to find out about them so that *when you make a presentation* you can address the misconceptions.

Novice salespeople sometimes believe that it is their duty to tell an advertiser that ads in the newspaper won't work or aren't appropriate. Marketing 101 tells you that you shouldn't go head-to-head with someone who got there first—because you'll lose. In other words, if an advertiser is spending 50% of her budget with the newspaper, don't tell her not to do that because it doesn't make sense. Without knowing it, you will cause the prospect to defend her actions. You'll be setting up the scenario of good cop/bad cop. You'll be the bad cop and the newspaper will become the good cop. Instead, you can study the company's newspaper buys. During your presentation, you can explain "I believe you can make your newspaper buys more effective. Here's a way you can change your newspaper ad and integrate it with your radio ad." Now, you've increased your credibility with your prospect instead of diminishing it.

Yellow Pages. Many advertisers don't think of their yellow pages advertising as a part of their "ad budget." A reality check discloses that "yellow page" companies are taking more money out of almost *every* market than any other medium. You can help your advertisers merge their ad expenditures in their minds and on their books while you help them analyze the effectiveness of their yellow pages advertising.

Radio. Almost any radio professional will tell you that radio people need to quit competing with radio people and concentrate on taking ad dollars away from other media. In the real world, however, advertisers often feel that they must make a choice between radio stations because their budgets usually do not allow them to buy on all the stations in a market—even if they all covered the same demographic group of consumers. Sometimes you *know* that an advertiser is mismatched with a particular radio station. You can't march into the business and declare that his advertising acumen is less than what it could be. At least, if you want to be successful, you won't. Instead, ask "Who is hearing your ads on that station?" Once again, you're going to find out what other media reps are telling the advertiser about their station. If you uncover major misconceptions, show evidence in your presentation that substantiates *your* profile of that station's listeners. Remember, every advertising medium has merit. Try to focus on the best way for your client to utilize ad dollars for the greatest return on investment.

Your Station. Perhaps the businessperson you're interviewing has advertised with your station in the past or makes small buys infrequently. You believe he has upgrade potential. Ask him "Who is hearing your ads on my radio station?" You may discover Grand Canyon–sized misconceptions about your station or its programming. Please avoid the inclination to discuss those misconceptions at this interview. It will be tempting, but it is imperative for you to keep this interview strictly an information-gathering meeting. If you don't, you may find that your prospect will clam up tighter than a federal building on a holiday. And, you'll lose your credibility.

Direct Mail. Direct mail has become a prime competitor for radio stations in the past few years and it continues to grow. Radio and direct mail can be very effective when used in tandem. Sales reps, however, need to increase their understanding of how advertisers are using it. As you know, one of the greatest disadvantages of direct mail is keeping the mailing list current. As soon as an advertiser receives a mailing list, it's ancient history. It's similar to the phone book. As soon as it's published, it's out of date. Many of the people listed have moved, businesses have closed, and new businesses have opened. Perhaps you can help your advertisers with direct mail's largest handicap. If the programming people at your station compile an ongoing database of listener information, you may offer to lease your "perpetually" updated list with a radio schedule. By combining the two media and using up-to-the-minute mailing lists, you'll be helping advertisers immensely and increasing revenues at your station.

Promotional Schedule

When you examine an advertiser's promotional schedule, you learn about any major promotions they feature during the year. As you discover what has been the most successful for this advertiser, you'll gain more insights about helping them become even more prosperous.

Sales Cycle

Every business has some months that have stronger sales than others. All businesses have some months that are weaker in sales. Too often, prospects and clients see advertising sales reps as "the media" and forget that radio is a business also. It's OK to explain to an advertiser that your radio station is a business with stronger and weaker sales months—just like his. It's easier for your prospect if you begin with a positive, so ask about his strongest sales month first. He's proud of his strongest sales month. It represents success. Then, ask about his weakest month. Next, ask which months are "fair" or "good" months. Sometimes you'll need to ask month by month to help your advertiser jog his memory. At this point, prospects have been known to take out a profit-and-loss (P & L) statement in order to give you accurate information. This is a solid indication that the prospect is interested in working with you. Consider it a compliment.

Compare. Once you've completed the promotional schedule and the sales cycle, compare them. An advertiser's promotional schedule usually mirrors the sales index because, when a business is promoting, sales are usually stronger. You may choose to point out the sales cycle and promotion connection during your presentation at the next meeting.

Closing Questions

When you ask a client what he would like from media salespeople that he's not getting now, you'll gain incredible insights. Sherry, a niche marketing sales rep, was interviewing Tim, the general manager of a regional department store chain. When she asked him what he would like from media reps that he was not getting now, Tim exploded with information, just like Mt. St. Helens at its peak. He said, "Do you know that I could put every media rep in this market in front of my desk. I could line them up side by side and put dividers between them. Do you know that they would all be saying the *same* thing to me! They would all be telling me that they're number one." "How can that be?" he asked. Without waiting for a reply, he continued, "And, I don't really care! I want to know how to sell my merchandise out there on the floor!" As he took a breath, he asked, "Do you know what they're *not* telling me? When one rep says that the station is number one, she doesn't tell me it's on Saturdays between 11:00 A.M. and 12:00 noon with young men 18–24! The next rep tells me that their station has senior citizens on Sunday mornings but doesn't tell me that's the *only* time it's number one! I don't *need* this information! Why are they doing this to me?" he pleaded.

Tim explained to Sherry that he understood that the reps were only trying to make a sale and wanted to present their station in the best possible light. His frustration level knew no bounds, however. His response to these robotic media reps was that he had given up seeing them face to face if he could avoid it. Sherry immediately reassured Tim that when she returned with her niche marketing presentation its focus would not be on being number one. "My focus will be on your business and on your customers," she promised. Because no one else had ever been in to see him and ask questions that seemed to genuinely concentrate on customers and marketing, Tim admitted that he was curious to hear what Sherry would show him in her presentation.

Next Appointment

The last question on the NMA: I form is not written in stone. As you can see, it says that you need about seven working days to complete your presentation. You don't have to ask for seven days to prepare. One of the goals in your life, however, may be to eliminate as much stress as possible. Conducting an interview today, attempting to prepare the presentation tomorrow, and giving it the next day—when you have eight other clients that need your attention—could put unnecessary pressure on you. You'll need time to read part of the trade magazine you borrow. You may want to check the Radio Advertising Bureau (RAB) marketing files about this type of business. Or, you may call the RAB research department to see what new information they may have compiled recently. You'll definitely want to study the answers to the NMA: I questions. You'll need to analyze the challenges your advertiser is facing so you can start thinking about the best ways to help. You may also need time to sit down with your copywriter and discuss your prospect's Positioning Analysis and Customer Profile.

Credibility. Once the interview is completed, many sales reps have had advertisers ask to make a copy of the interview form for their records because they've never had so much of their advertising and marketing information on one piece of paper. Some prospects have shared with salespeople that their spouses don't even know as much about their business. "In fact," they confide, "no one has ever spent as much time to learn about my advertising needs." Imagine how much credibility you will gain and how it will make you stand out in an ever-increasing field of media robots.

Customizing. The NMA: I form covers magazines, direct mail, billboards, cable and over-the-air television, radio, yellow pages, inserts, and newspapers. Perhaps magazines, for example, aren't major competitors in your market. In some markets, billboard companies are large competitors; in other markets, they aren't. You can customize the NMA: I for your market by leaving out the questions about the media that doesn't represent your main competitors. You'll have more room for notes or you may want to add questions about co-op advertising.

Choosing the Marketing Path

The answers to the Niche Marketing Analysis: Part I questions will give you the general direction, or marketing path, your client will find most beneficial. Your presentation will outline a series of campaigns to be created over a period of time. Now you have information about your prospect's customers and you know a little about the company's buying habits. You have a sketch of the advertiser's media use and her perceptions of other media. You have the names of the decision makers. You have a glimpse of advertising goals. You have information about the prospect's promotional schedule and sales cycle. You're now ready to prepare a presentation.

Congratulations

You've now opened the door to establishing a clear-channel communication system with your advertiser. Next, you'll need to compile and present your information. Chapters 10 through 14 will give you an in-depth look at creating and giving niche presentations. Next, let's look at some innovative ways to develop your databases and make more money.

5

Database Marketing

Learning the Tush-Push

Today, if you buy a video that teaches you how to do the Tush-Push, within a few weeks you'll get an offer in the mail for more videos teaching the Cotton-Eyed Joe, line dancing, or the two-step. Some video retailer has purchased information about you and your buying pattern for videos—especially the fact that you've recently acquired the Tush-Push dance training video. Whether you know it or not—whether you want to be or not—you are a part of the database marketing world. Cutting-edge businesses are using database marketing to find the best prospects for their products and services. They know that it's less expensive and more effective to reach their "best" prospects rather than blindly sending their message to thousands of general-market consumers. Companies such as Levi Strauss, Cannon, and Saab are diverting large portions of their media dollars to database marketing projects.

While the cost of media escalates, the cost of storing information on computer disk is plummeting. In the 1970s, storing names on disk cost more than $7 per page. Today, the cost is less than one cent! Saab, for instance, sent a direct-mail piece to their existing customer base asking if owners would like to see a video about specific new models. The video came with a reply card asking for more information on the customer's views about Saab (Figure 5.1). The marketing strategy is basic niching: develop share-of-customer instead of share-of-market.

Shrinking Media Budgets

Whether or not you decide to participate in this mega-marketing trend, you might want to know some of the principal highlights of database marketing. Then, you'll understand where your media dollars have gone when media budgets are cut—again. You also may want to participate in database marketing to gain additional revenue at the ad agency and local retail level. Retailers, service companies, and ad agencies like to use current databases because they connect with consumers on a one-to-one basis in nearly

You've seen the film, now give us your review.

We'd like to know what you think of this Saab video. So please take a moment to complete the following questions.

1. Was the video sufficiently informative regarding:
   Safety features? ☐ Yes ☐ No   Performance features? ☐ Yes ☐ No
   Environmental features? ☐ Yes ☐ No   Comfort/practical features? ☐ Yes ☐ No
2. Would you be interested in receiving additional videos on the Saab product line? ☐ Yes ☐ No
3. After viewing the video, are you more likely to test-drive a new Saab? ☐ Yes ☐ No
4. Would you like us to send this video to a friend? Please provide mailing information below:
   Name (Mr./Mrs./Ms.) _____
   Address _____ City _____ State _____ Zip _____
   Telephone ( ) _____

Please fill in your name and address:
Name (Mr./Mrs./Ms.) _____
Address _____ City _____ State _____ Zip _____
Telephone ( ) _____

Thank you for your time.

FIGURE 5.1 Saab is building more customer information for a database.

real-time frames. The downside for radio stations is that maintaining a current database takes a personnel commitment and a serious financial outlay, which may run as high as $18,000 to $50,000 per year, depending on the number of outside sources who supply information to your line. For instance, you may want to hire a comedian to provide a laugh line or a weather service to provide long-range forecasts. The good news is that current staffers, such as your news director, can provide some information. An announcer can maintain a concert line, for example. The up side for radio stations is that a well-maintained, up-to-the-minute database can pay for itself plus generate added revenues to your radio station.

Mountains of Information

Some radio stations have thousands of listener names in their computers. Other stations include the names and addresses of advertisers. Cutting-edge radio stations are now including information about their advertisers' customers. There is, however, no clear-cut format for using all the information, but some suggestions on how this new resource can be used for additional sales can be found in Chapters 10 and 11. According to Mark Heiden and Linda Brown of Eagle Marketing, Inc., of Fort Collins, Colorado, there are three ways to look at your database—"Your options are: (1) demographics, which is circumstance; (2) psychographics, which is attitude and behavior; (3) geography, which is location."

The overriding consideration in database marketing is timeliness. The information needs constant updating. Information that's more than six months old will probably not be accurate. You may wind up with a collection of names at old addresses but not much else. To keep the database fresh, a full-time manager can update data on a daily basis. The basic information you may want to retain is: listener names, addresses, phone numbers, interests, lifestyles, preferences for goods and services, and most recent purchases, as well as listening data. Be selective about the offers you make, however. Study your advertising customers and your listener customers carefully. You have a responsibility to protect your database to ensure its longevity and effectiveness. Making offers that a consumer would be interested in receiving is the key. A prospective new home buyer is only interested in new home information until they make a final purchase; then, new home information becomes a bother—even a nuisance.

Database Goals

Developing a database, by itself, does not provide a solution to anything. You need the ability to communicate with individual listeners. Databases are only a means to a specific goal. The goal may be direct contact with listeners, or the goal may be dialogues with consumers that tell you more about what your advertisers' consumers want and need. The result may be increased sales for your advertisers or more time spent listening to your radio station or both. In other words, you're trying to get closer to your customers—whether you're studying them as listeners or consumers of your advertisers' goods and services.

The Computer Edge

A competent computer operator can run your database and perform telephone call-outs. Keeping data fresh is critical. Scanners and bar codes make the processing of information quite efficient in retail stores and remote locations. Powerful comparison studies of retail clients can be run; for example, you can call up a list of all married women in particular postal zip codes or block groups. You might want to show a prospective advertiser how many of the 25–34-year-old listeners in your database are married, single parents, or engaged, because marital status affects purchasing habits. High-volume retailers, like fast-food restaurants, love these listener insights. The information helps them target niches and helps create niched campaigns to move merchandise. A database with 50,000 listener members is a far more valuable resource than the standard mass-market ratings service audience samples.

Know Your Listener Consumer

Study your listeners until they are as familiar as the back of your hand. What is their family makeup, including educational level of all family members; do they have children, and even pets. What is their purchasing history? What have they purchased in the past that matches your advertisers' products and services? What is their purchasing ability—such as combined income, the value of their home, their assets, and the credit cards they use? What are the occupations of *all* household members? How do they feel about receiving direct-mail and/or telemarketing offers or buyer's clubs? What do they do in their leisure time? What are their hobbies? Are the listeners sports fans? To what degree? Do they read specialty magazines or newspapers? What other radio stations do they enjoy? Which television programs do they watch? When you have compiled this information, you've begun a basic database relationship. Don't become overwhelmed—this isn't as difficult as it appears at first.

Interactive Options

One of the elementary ways database marketing is being used to create listener relationships is with the Universal Product Code (UPC). For example, you can create an interactive listener telephone club. Distribute plastic cards, similar to credit cards, that store and identify listeners by using bar codes. Individual station memberships run as high as 100,000 people. Tied in with interactive phone services, members can sample CDs, get their horoscopes, participate in station polls, and receive discounts on specific products at participating supermarkets and other retail advertiser outlets.

Kraft Foods uses the WKYS Club (the KYS Connection) in Washington, DC, to provide discounts at stores such as Safeway. The bar-coded club card becomes the coupon, which saves money on printing and coupon distribution. The cards are scanned at the checkout counter and the information is stored on computer. Specific purchases by individuals can be tracked and (if the database is current) a great deal can be learned about customers (listeners). The purchasing information is available in almost real-time and provides a listener link that can attract more advertising investments for your radio station (see Exhibit 5.1). Services like the KYS Connection are supported by regular direct mailings as well as systematic call-outs to members' homes; this kind of interactive club can be a valuable profit center for radio stations.

Variations

There are variations of the KYS Connection that you may want to sample. KLOL in Houston, Texas, offers the Concert Hotline—(713) 266-2378. It's

Exhibit 5.1 A Sample Rate Card

THE KYS CONNECTION

Sponsorship Availability

The KYS Connection receives over 60,000 calls monthly from WKYS-FM listeners. Several of the featured lines are available for sponsorship and client participation.

Featured Line	Monthly Investment
Weekly Horoscope	$1,200
CD Sample Line	600
Cla'ence Update	575
Laugh Line	300
Weather	300
Cosmic Forecast	300
Concert Hotline	225
Cultural Events	100
Sports Update	100
Children's Activities	75

Lines are available for special interactive telemarketing uses.
Sponsorship includes: Weekly promos PLUS a Sell-line on the KYS Connection.

---

more than just a concert information line. You can call and listen to information—local weather, enter station contests, or become a lifetime member—just by hitting a combination of letters on a touch-tone phone. WZZO in Allentown, Pennsylvania, offers an Information Line—(215) 435-9550. The line enjoys sponsors, such as Budweiser, and offers a variety of information services. One of the best features of these interactive services is the ability to be transferred directly to a retailer. For instance, if you listen to a sample of a CD that you like, you can touch a number on your phone and instantly be connected to a retailer who sells the CD. The KYS Connection allows a caller to have direct access to retailers; using the WIZ Sample Line, Moviefone, or auto leasing line, callers have the option of direct access to retailers after they hear the information requested.

Added Value. To give added value, you may also want to include postcard mailings with sponsor logos to help promote the line. These sponsor cards may be used as additional handouts at remotes and mail stuffers, and to enhance client presentations.

Brite Voice Systems. According to Carolyn Russell of Brite Voice Systems in Wichita, Kansas, there are very few limits on the kind of information you can offer. Currently, Brite Voice serves over 2,000 clients worldwide, offering more than 150 information programs. These programs are distributed via satellite to interactive phone information lines. The lines are offered by organizations such as the media, health-care facilities, and companies that offer financial services.

Database Wars

As you identify and store more information about listeners, advertisers, and advertisers' customers, database wars are inevitable. It happened with the airlines. It's happening with the credit card industry. People with good credit histories are being bombarded by competitive companies offering lower interest rates. Database wars will happen with radio listener clubs in the future. The battle lines were drawn by airline frequent-flyer clubs and hotel/motel preferred-customer clubs that developed in the 1970s and 1980s. The bonus miles and points accumulated by members give marketers a direct line to their best customers. These databases also yield information about nonconsumers (nonlisteners). They become the target of direct-mail and phone campaigns to attract them to your radio station. Your computer ingests reams of information about individuals that can be cross-referenced and used for niche sales opportunities. Advertisers are dying for this kind of information. You don't have to cut rates when you're offering advertising packaged with timely databased information and related promotions.

The Price Trap

Giving away free miles, free rooms, or concert tickets to participating club members helps build loyalty and current data. You may even want to consider charging a nominal fee for joining your club—like the Holiday Inn Priority Club and American Express Gold Card Mileage Plus programs do. The trap some companies fall into is lowering rates for their core customers. Pan American and Eastern Airlines, both now defunct, hastened their ultimate demise by engaging in unnecessary price cutting to members of their already-discounted frequent-flyer clubs during the 1970s airline deregulation era. Offering value-added items, such as special in-flight meals on full-price tickets or room upgrades when a hotel is underbooked, makes economic and marketing sense. Offering these same concessions to discount ticketholders or free upgrades when a hotel is already full is economic suicide. By comparison, in radio it makes economic and marketing sense to include a Weather Line card in mailings to people who fish commercially when you're charging premium rates to companies who sponsor your Weather Line. In this scenario, everyone wins.

Database Commitments

As company marketing managers move from the steamroller tactics of mass marketing, radio loses money to direct mail and telemarketing. To counter this revenue loss, you may want to learn the fundamentals of a product-driven individualized marketing program. The first rule is to identify prospects every day. Seagram's, for example, has a current database of over 7 million prospects and the list grows by approximately 20% every year. A good direct-marketing manager will try and involve everyone in the company in the business of identifying prospects and customers (listeners). Advertising, packaging, event marketing, and in-store promoting become a means for building your database. Radio managers who think like direct-marketing managers will enjoy more success. Information from people who visit remotes, complete contest entry forms, win free dinners, attend trade shows, and respond to contest lines should be a part of your database. It is also important to have people who do not listen in your database. Discovering which stations people listen to can help you draw comparisons to your core listener base. Then, if you're interested, make programming, marketing, and promotional adjustments to see if these new people are worth attracting.

Restaurant Marketing Opportunities

One of the best year-round client categories for radio stations is restaurants. They love to promote menu items, and free dinners make great prizes for radio station listeners. Many restaurants are also notorious for not keeping records. Your station can help restaurant clients by creating detailed information forms for winners that can help the restaurant better understand its customer base (see Exhibit 5.2). Other than the obvious name, address, and phone number, you also can track the amount won and the additional amount spent—usually more than the amount won—by the winners. Combine this survey form with a 26-week schedule on your radio station. Offer to tabulate the results of the study, and build the price of that into the package.

The Information Advantage

Entry forms can give your client a focused look at the market. Your radio station will have a competitive advantage that other media do not enjoy. This type of activity is an excellent form of relationship marketing. You enjoy an information advantage in an age of information.

Exhibit 5.2   Restaurant Registration Form

REGISTRATION (*Insert Station Logo*)

PLEASE COMPLETE THIS FORM AND DEPOSIT HERE.
ENTER AS OFTEN AS YOU VISIT. YOU MUST BE 18 TO ENTER.

☐ First visit   ☐ Second visit   ☐ Third visit   ☐ Too many to count

I/We had: ☐ Breakfast   ☐ Lunch   ☐ Dinner   ☐ Buffet   ☐ Menu

There were _____ people in our party.

I came with ☐ Family   ☐ Spouse   ☐ Friend(s)   ☐ Business associate(s)
☐ Other: _____

I/We ordered: ☐ Appetizers   ☐ Chicken   ☐ Seafood   ☐ Steak   ☐ Specials
☐ Salad   ☐ Dessert   ☐ Other: _____

The items I/we liked best were: _____

I/We didn't like: _____

because: _____

I/We rate your food as:	Excellent	Very Good	Fair	Poor
Presentation	☐	☐	☐	☐
Flavor	☐	☐	☐	☐
Portion size	☐	☐	☐	☐
Correct temperature	☐	☐	☐	☐
Pricing	☐	☐	☐	☐

I/We rate your service as:	Excellent	Very Good	Fair	Poor
Friendly	☐	☐	☐	☐
Prompt	☐	☐	☐	☐
Attentive	☐	☐	☐	☐
Accurate	☐	☐	☐	☐

How often do you visit Willowbys?   ☐ Daily   ☐ Weekly   ☐ Monthly
☐ Yearly   ☐ Other: _____

What brings you back to Willowbys?   ☐ Price   ☐ Atmosphere   ☐ Service
☐ Generous portions   ☐ Other: _____

How far do you travel to visit Willowbys?   ☐ Less than 1 mile   ☐ 1–5 miles
☐ 6–10 miles   ☐ 11–15 miles   ☐ 16 miles or more

The decision to eat out today was: ☐ Made at the last minute
☐ Made 1–2 hours ago   ☐ 3–4 hours ago   ☐ Planned 1 day ago
☐ Planned a week ago   ☐ Other: _____

Are you: ☐ Vacationer   ☐ On business   ☐ Living nearby
☐ Passing through   ☐ Other: _____

Your age range: ☐ 18–24   ☐ 25–34   ☐ 35–44   ☐ 45–54   ☐ 55–64
☐ 65–74   ☐ 75+

Name: _____   Phone: _____

Street address: _____

City: _____   State: _____   Zip: _____

Department Store Marketing Opportunities

According to the Radio Advertising Bureau, November and December are the best sales months for department stores. The marketing challenge is to find ways to build business in the 10 slower months. Radio stations can help their department store advertisers by adding to their local retail customer information. Go directly to the individual retail customer with a questionnaire. (Methods for presenting database-building forms to advertisers can be found in Chapter 10.) The questionnaire in Exhibit 5.3 helps build the facts needed to create an up-to-the-minute database for local department stores.

Sample Base

Researchers like to use sample bases of at least 300—or more—to get an accurate picture of a demographic group. Building an information base of customers in specific demographic cells, such as men 35–44, will help a department store gain the knowledge it needs to increase sales. Understanding the purchase cycles of a particular product, such as socks, helps stores target customers when they're ready to replace a product. The axiom in political circles also applies to retailers and people who sell services: "Savvy politicians start planning their next election 24 hours after they're elected to office." Your advertisers *continually* need to be collecting data from consumers.

Advantages of Database Marketing

The Image Trap

Some advertisers will be hesitant to use database marketing. They've heard and seen elaborate image campaigns such as Budweiser's "This Bud's for You" and "Proud to be Your Bud." In the general market arena, a good example is the yellow pages campaign: "Let your fingers do the walking." Coca-Cola used a mass-marketing, steamroller technique when they introduced Coca-Cola in Europe after World War II. Europeans did not react to the famous red button that said, "Drink Coca-Cola." In the late 1940s, Coke cost about $1 for a 6-ounce bottle; Western Europeans could buy a liter of wine for about five cents.

The image, or reminder, advertising was not effective with Europeans. The pre-sold American G.I., however, helped make Coke a success in Europe. In Europe today, even fast-food outlets, such as McDonalds and Burger King, sell wine and beer for about the same price as a soft drink. In this marketing situation, image and reminder campaigns did not produce the share-of-customer that database marketing can produce. The majority of the population had been preconditioned to drink wine or beer with lunch and dinner. Five decades later, most Western Europeans still

Exhibit 5.3 Department Store Questionnaire

REGISTRATION *(Your Station's Logo)*

PLEASE COMPLETE THIS FORM AND DEPOSIT HERE. ENTER AS OFTEN AS YOU VISIT THIS STORE. YOU MUST BE 18 TO ENTER.

This is my: ☐ 1st visit ☐ 2nd visit ☐ 3rd visit ☐ Too many to count

I visit _____ : ☐ Daily ☐ Weekly ☐ Monthly
☐ Other: _____

I have shopped in: ☐ Kids' department ☐ Men's ☐ Women's ☐ Gifts ☐ Housewares ☐ Jewelry ☐ Shoes

Today, I found the merchandise I wanted: ☐ Yes ☐ No
If not, the reason was: ☐ Selection ☐ Price ☐ Quality
☐ Other: _____

What I like most about Michaels: _____

What I like least about Michaels: _____

Sales associates are:	Yes	No
Prompt in waiting on me	☐	☐
Friendly and helpful	☐	☐
Knowledgeable about the merchandise	☐	☐

My birthday: _____ My anniversary day: _____

Service at Michaels is: ☐ Outstanding ☐ Satisfactory
☐ Not satisfactory ☐ Other: _____

My favorite day(s) to shop are: _____ Hours: _____

I use: ☐ MC ☐ Visa ☐ AmEx ☐ Discover ☐ Diners Club
☐ Cash ☐ Check

I like to shop: ☐ Specialty stores ☐ Catalogs ☐ TV shopping
☐ Factory stores ☐ Other: _____

Rate (use 1 for highest) in order of importance: _____ Selection _____ Price
_____ Service _____ Convenient hours
_____ Other: _____

I have used Michaels: ☐ Gift wrap ☐ Packing/Shipping
☐ Free alterations ☐ Personal shopper

Age range: ☐ 18–24 ☐ 25–34 ☐ 35–44 ☐ 45–54 ☐ 55–64
☐ 65–74 ☐ 75+

Name: _____ Phone: _____

Street address: _____

City: _____ State: _____ Zip: _____

respond to their childhood beverage conditioning. Database marketing may help advertisers find new niches for selling alternative beverages.

Cigarette Lessons

Radio was considered so powerful by the U.S. Congress that it banned on-air cigarette advertising in 1972—right after the Super Bowl. During the 1940s, cigarette companies did their patriotic duty by giving free product to members of the armed services. After the ban on cigarette advertising, cigarette companies reverted to elaborate questionnaires asking consumers all kinds of questions about their consumption habits, including cigarette consumption. In the 1980s, companies like R.J. Reynolds invested in excess of $100 million to build its database and generated cigarette information for about 25 million Americans.

Today, two decades after the ad ban, cigarette companies still use database marketing as a powerful tool (see Figure 5.2) because this marketing strategy is so effective. The information sought includes cigarette consumption, information about chewing tobacco, and beer consumption. The additional information may be sold to other companies to defray the cost of generating database information. There are mass-media options (for example, billboards and newspapers); and niched media options (such as specialty magazines). Yet, media receives diminishing budgets from the cigarette industry in favor of databased marketing budgets. The marketing message is loud and clear. The more you understand the processes of databased marketing, the more you'll be able to take advantage of these new sales, marketing, and listener-building opportunities.

The Economics

Most seasoned radio managers were raised in the era of mass marketing. Increasing your audience share by one point in a major market could mean a million or more dollars in additional revenue. One only had to employ salespeople who had a facility with numbers. Many radio managers aren't used to dealing with listeners or customers on a one-to-one basis. The strategies of mass numbers drive radio people to invest dollars in programming instead of database marketing. Thanks to continued market fragmentation, however, audience numbers will continue to shrink. The economics of mass-audience marketing will make less sense. Most managers are aware of fragmentation, but marketing to individuals looks expensive. The popular 25–54 demo provides a comfort zone for many radio managers, agencies and retail clients. What does a 25-year-old have in common with a 54-year-old? Not much. Still, management continues to cater to the needs and wants of their 25–54-year-old customers. This demo is a marketing illusion.

A more cost-efficient strategy is to engage in one-to-one marketing with the goal of uncovering potential core listeners. These are the loyal

FREE COUPONS FOR SMOKERS AND TOBACCO USERS

Complete this American Smokers Registry survey to become eligible to receive coupons for fine tobacco products.

0 0 1 8

Date: __________

Name #1 __________

Address __________

City __________ ST ______

Zip ______ Phone (___) __________

Year Born 19 ____ Sex: ☐ M ☐ F

CIGARETTES:

1st Choice Brand __________

Type: ☐ Light ☐ Ultra-Lt. ☐ Other
Flavor: ☐ Menthol ☐ Non-Menthol
Size: ☐ King/Reg. ☐ 100's ☐ 120's

I Usually Buy: ☐ Packs ☐ Cartons

2nd Choice Brand __________

Dip or Chew? ☐ Yes ☐ No

Brand: __________

Favorite Beer: __________

2nd Choice Beer: __________

I Usually Buy: ☐ 6Pk. ☐ 12Pk. ☐ Case ☐ Other

Signature: __________
(Required)

Name #2 __________

Address __________

City __________ ST ______

Zip ______ Phone (___) __________

Year Born 19 ____ Sex: ☐ M ☐ F

CIGARETTES:

1st Choice Brand __________

Type: ☐ Light ☐ Ultra-Lt. ☐ Other
Flavor: ☐ Menthol ☐ Non-Menthol
Size: ☐ King/Reg. ☐ 100's ☐ 120's

I Usually Buy: ☐ Packs ☐ Cartons

2nd Choice Brand __________

Dip or Chew? ☐ Yes ☐ No

Brand: __________

Favorite Beer: __________

2nd Choice Beer: __________

I Usually Buy: ☐ 6Pk. ☐ 12Pk. ☐ Case ☐ Other

Signature: __________
(Required)

Name #3 __________

Address __________

City __________ ST ______

Zip ______ Phone (___) __________

Year Born 19 ____ Sex: ☐ M ☐ F

CIGARETTES:

1st Choice Brand __________

Type: ☐ Light ☐ Ultra-Lt. ☐ Other
Flavor: ☐ Menthol ☐ Non-Menthol
Size: ☐ King/Reg. ☐ 100's ☐ 120's

I Usually Buy: ☐ Packs ☐ Cartons

2nd Choice Brand __________

Dip or Chew? ☐ Yes ☐ No

Brand: __________

Favorite Beer: __________

2nd Choice Beer: __________

I Usually Buy: ☐ 6Pk. ☐ 12Pk. ☐ Case ☐ Other

Signature: __________
(Required)

I certify that I am a smoker; that I am 21 years of age or older; and that I want to receive offers, premiums, coupons or free cigarettes that may be sent to me in the mail. I understand that giving false information in order to accept these offers may constitute a violation of law.

FIGURE 5.2 Offers for free coupons help generate information.

people who attend remotes and fill survey forms with large blocks of time spent listening (TSL). These people are found by using database-marketing techniques. Such loyal listeners are also valuable to your advertisers because a marketing database builds repeat business. Marketers are now studying the *lifetime value* of a customer rather than how *many* customers can be persuaded to go into a business just once. A repeat sale costs about half of what the first sale costs. Increasing repeat sales through database marketing can bring total costs down and profits up.

Leasing Your Database

Once you've built your database, you may want to consider one of the newest sources of radio revenue—database leasing. There are several issues to consider: privacy and station/listener relations. To avoid problems, you should retain control of the listed information at your station. According to Mark Heiden and Linda Brown of Eagle Marketing, Inc., there are important guidelines for database leasing. First, you need to define usage rights. Will this be a one-time usage, multiple usage, or an annual usage? Second, will the client lease all the information in the database or a subset of it? For example, a client might lease only geographic and lifestyle information from the database. The inherent value of predisposed recipients could be from $25 to $60 per thousand—or higher—based on the quality of the information you have in your database. You can obtain a form from U.S. post offices (P.O. Form 3602) that is an acceptable proof of performance for co-op or clients; it verifies how many pieces you mailed and when.

CPP—A New Definition

CPP traditionally has represented *cost per point*, or the cost of a radio ad measured against 1% of the population being measured. This is a mass-market approach. Traditional CPP definitions relate more to mass-market television. The new definition for CPP in niche and one-to-one marketing is *cost per prospect*, which is a much more realistic measure of the effectiveness of a radio schedule that targets niche audiences. For example, let's assume that a mass-market $100 television ad reaches 100,000 persons who are 25–44 years of age according to the projected audience estimates. Let's also assume that a $20 radio ad reaches 1,000 persons, aged 25–44. On the surface, the television schedule is a clear winner. Why buy radio? Enter database marketing.

Now, you're able to show that 50 of those radio listeners are prospects for a particular type of merchandise or advertised service, but only 10 television viewers are in the market for the same goods and services. Thanks to your database, radio receives the order. Cost per prospect, thanks to databases, is a viable alternative to the way media has traditionally been purchased. Radio station databases with current listener information provide an efficient, economical avenue to access this vital sales and marketing information.

A Fresh Start

Throughout this book, you'll discover new ways of developing revenue streams for radio stations. Database marketing can provide interactive listener responses, which can be used to sell more advertising. Listeners can be connected directly with retailers through interactive phone systems. Direct-mail advertising opportunities are available to specific listener groups in your database. Cost per prospect information—the new CPP—can be used at ad agencies to develop new business for radio stations. Niche marketing opportunities can be uncovered in your database. And, programmers can develop one-to-one listener dialogues with core listener groups. In a few years, database marketing will become so much a part of our everyday business life and culture that it will be hard to imagine life without it.

Next, let's examine some of the opportunities focus group research can bring to your radio operations.

6
Focusing on Focus Groups

Should You Consider Focus Group Research?

A journey into the world of focus group research may seem as simple as surviving alone for a month in the middle of a jungle. After all, you're in the radio business—not in the research business. Yet, everywhere you look, you see a world learning about itself through interactive research. Focus groups can tell you a great deal about what motivates people to listen to your radio station, and they can help you uncover new programming niches. Focus groups can also provide management with a one-to-one look at how people perceive their radio station. Properly executed, focus groups can reveal four things that are critical to your radio station's programming success:

1. They show you *what* your listeners want.
2. They help you find out *when* listeners want what they want.
3. They show you *how* your listeners want what they want.
4. They let you see *why* listeners want what they want.

Not understanding one or more of these four critical areas is like trying to sit on a three-legged stool with one or more legs missing. You're bound to fall over.

Until recently, focus groups have been used for many purposes, including testing ad copy, improving station marketing themes, and learning listener attitudes about a station's programming and personalities. Qualitative research of this kind uncovers in-depth views and insights that are impossible to detect in nose-counting, quantitative research. Focus group interviews encourage panelist interactions that help expose underlying motives that prompt people to listen to your radio station and buy your advertisers' goods and services. But, focus group research isn't just a programming tool.

Focus group studies provide sales information for your sales department and marketing people. The reports from focus group interviews can help your salespeople profile your listeners more clearly for business decision makers. The information can enhance sales presentations by demonstrating the strengths of your programming niche in terms that will help clients move goods and services. If your advertisers and your salespeople also want to know what consumers want from your clients' businesses, why not ask them face-to-face? Focus groups can tell you a great deal about what motivates consumers to buy from your advertisers. These effective techniques are demonstrated in the presentation found in Chapter 14.

An Opportunity

In the past, focus group research was conducted only by research firms who specialized in qualitative analyses from focused-interview studies and ad agencies that had facilities to accommodate focus groups. Focus group research was understood by the layperson about as much as listeners understand the inner workings of a radio station. Today, however, many businesses are conducting their own focus group studies. These forward-thinking companies are trying to reduce the risk factors in decision making. They have realized that their organizations will spend less money and increase their profits faster when they understand their customers and noncustomers better.

Top-of-mind awareness is a concept used often by radio salespeople to explain the need for effective frequency in a radio schedule. Focus group research delves below the surface and uncovers motivations to purchase that are not *top-of-mind*. Understanding these impulses can help a company develop far more effective advertising and marketing strategies, better customer services, and a more-profitable product line. Tomorrow, the most successful radio companies will offer focus group research to their advertising clients in the continuing quest to become true marketing partners.

Benefits?

You may want to commission a focus group study to learn more about your radio station's listeners. And, you may decide to offer an additional service to your advertisers—focus group research. Let's explore focus groups from the perspective of a customer (you) purchasing a focus group study to learn more about your listeners and nonlisteners. Then, let's explore the process for conducting focus group interviews for your station's advertising clients. When you become the facilitator of focus group studies, you'll be increasing your radio station's revenue stream while offering a service that your competitors do not offer. You'll also be strengthening your marketing partnerships with your clients. Eventually, you'll work with fewer clients because you'll be getting more business from each client. You'll be increasing your share of customer rather than your share of market.

Commissioning the Study

When you commission a focus group study for your radio station, you'll be asked to help shape the questions for the panelists. Your input may also be requested when decisions are made about the direction the research will take. The moderator will want to thoroughly examine your goals for the project. Be specific about what you think the research will tell you. The more you understand about the facilitator's needs and abilities, the more you stand to gain as a client. Successful focus group interviews require a multitude of skills. First, you need the leadership of well-trained facilitators (group leaders). Look for a facilitator who has some knowledge of the radio and/or media business.

Discussing Price

When you interview a candidate to conduct your focus group study, after you have discussed what you want to achieve from the research, ask how much they will charge and what that amount will cover. Some focus group companies will offer "complete packages" and some will work "a la carte." More than one set (three or four different sessions) of interviews may be necessary if your radio station serves a market that covers a very large geographical area or if the market contains a mosaic of ethnic groups (most markets fall into this category). You'll need an adequate sample base of your station's listeners drawn from your hot zip codes, hot blocks, or your core audience. Or, you'll need to interview a group of people who represent the core audience of your major competitors—your nonlisteners.

Hiring the Right Firm

You may ask for a proposal from each of the research candidates. It probably will be no longer than a page or two stating the general problems to be examined, the goals for the project, and the expected cost. If you would like a longer proposal, expect to pay for the time the firm invests in formulating an estimate. After reading the proposal, do you feel the researcher *understands* your goals? Did she (most focus group moderators are female) outline *her* responsibilities and *your* responsibilities? You'll be working closely with the moderator of the interviews. This person may also be the report writer. She must gain credibility with you in order for her findings to appear valid. Simply stated, you need to *like* her and *trust* her.

Invest time on the phone and in person, if possible, finding out if you are *simpatico*. From your perspective, if a candidate is overbearing during the interview process, how will they appear to panelists? Does this person *look* like your listeners? Will participants in the focus group interviews feel comfortable with this person? Does she have a neat appearance? Do you sense that the prospective moderator *understands* your needs? Answer these questions thoughtfully; trust your instincts. Then, to demonstrate your professionalism, make your hiring decision in a timely fashion.

After making your determination, call the other candidates and explain why you didn't choose their firm. Avoid the tendency to tell the people with whom you didn't choose to work that you chose another firm just because their price was too high—unless that *was* the only reason. Explain why you feel the chosen firm fits your needs best. If you've spent any time in sales, you know how much this kind of consideration is appreciated.

Choosing the Site

You will need a carefully arranged room to conduct focus group research. To prevent possible bias from the group panel to be interviewed, this room must be off-site—not connected with your radio station or staff. The researchers need time—usually 4 to 6 weeks—to plan and attract participants to focus group sessions. A corporate conference room or a hotel meeting room may be sufficient. You will need, at least, an audio tape recorder. Preferably, you'll have a video recorder, too. If you, the client, want to view the proceedings, you will need to obtain an interview room equipped with a two-way mirror and an audio speaker system connecting the viewing room with the interview room. The advantage in viewing the session is that you can see the body language and the interaction of the participants much better through the mirror than anyone can catch it on videotape, regardless of the number of cameras in the interview room. When choosing a location, consider how far your participants must travel and what time of day the interviews will be conducted. The less time it takes a panelist to travel to an interview site, the more likely you are to have a full panel.

Bias

We live in a world of contradictions. The example of a young orphan trained to tell *the truth* by a religious order illustrates the point. Once lucky enough to be adopted, the youngster is taken by his new parents to a Sunday meal at the home of new relatives. As the front door opens, the youngster sees a very large female relative in an even larger yellow polka-dot dress. Immediately, the adoptee blurts out "What a big yucky yellow dress." The new father grips the child's neck firmly as a reprimand. The head on the big yellow polka-dot dress shakes slowly from side to side showing disapproval. As the family begins to dine, the new son is offered some freshly prepared peas from the host's garden. "I don't like peas!" exclaims the adoptee. This announcement is immediately followed by a pinch on his kneecap from his new mother, who is seated next to him. A quick survey of the room tells the youngster that his declaration is not well received. Soon, the child learns that *the truth* is not always in vogue. Many children learn at a very early age to often withhold their true feelings in favor of pleasing others. Bias is born.

Bias may be defined as a highly personal and totally unreasoned prejudice—about anything. Our young adoptee may grow up hating the color yellow or just disliking Sunday afternoon dinners. Attempting to sell him a yellow shirt or a yellow house may be equally unsuccessful. In the flurry of growing into an adult, he may not remember why he can't stand the color yellow but he gets a feeling of tightness in the back of his neck just thinking about it. In a focus group setting, bias may run rampant and needs to be uncovered. People do not always share their true feelings. Conditioned from childhood, they often forfeit the truth (as they see it) in favor of approval from others. Bias is a part of every human being and understanding the bias factor may help you get better results from your research investment.

Focus Group Composition

Unlike database marketing programs, which draw on relatively large listener samples—usually 10,000 or more listeners—focus groups are quite small. They tend to range from 8 to 12 people. A session usually lasts anywhere from one to two-and-a-half hours. The participants are usually the same sex and in a narrow age grouping (e.g., women, 25–34). Same-sex interviews tend to avoid the proclivity for male participants to dominate a session when the panel is composed of men and women.

Radio stations often use focus groups to test music with music hooks. These studies also may be used to test television ads for your radio station. Focus group research is quite useful in uncovering listener attitudes and perceptions about a particular station or format. For example, your moderator could play a generic hot country ad and then ask participants which radio station comes to mind. Panelists may also be queried about the content of the ad to determine if anyone found it fun, amusing, offensive, boring, or even stupid.

To gather a sufficient amount of data about a particular format or station perception, as many as 10 different groups of 12 people *may* be needed. Many focus group studies, however, use as few as 8 panelists and conduct anywhere from 4 to 8 interview sessions. To avoid bias when testing music, individual headsets help keep one or two people from dominating the group and forming a group opinion about individual music hooks. The group discussion may then dwell on formatics such as contests the listener may enjoy, a proposed positioning statement, and discussion about specific contest prizes.

Focused Prizes

Focus group research can help programmers and managers discover new ways to attract an audience. Let's look at a typical station prize scenario: The program director needs prizes for the station's prize closet. The sales manager needs increased billing. A salesperson sells a half-cash, half-trade contract to a jewelry store for $100,000. The station will receive $50,000 in new cash billing plus $50,000 in Rolex watches. Station ownership is ecstatic. They want Rolexes for their family to wear at country club events. The 100,000-watt station in the mid-South has a traditional country format even though it promotes itself as a *new country* station. Their core audience is persons 35–64 years of age. The audience mix is approximately 60% male and 40% female. Let's give away expensive watches. Right? Well, nobody consulted the listeners. Rolex watches are in vogue at the owner's country club; they are not a prize most desired by the two-step dancers at a (traditional) country music bar.

Focus group research can help identify the kinds of prizes listeners want. The simple questionnaire in Exhibit 6.1 was used in focus group research to query listeners about prizes that might make them tune into a country station. It is a brief example of what you might be called on to help develop as part of a listener focus group that examines the attractiveness of various types of listener prizes.

The Winner Is . . .   The clear winner among females who listened to a country competitor was not a new truck. It was an automobile. New trucks were second. Not surprisingly, a majority of the males in the focus group chose the truck. Free gasoline for a year was chosen third. Rolex watches were near the bottom of the desired prizes list. This station would do better to reward key employees and clients with the watches and give listeners a chance to win vehicles.

The Age Question

The age ranges listed in the questionnaire may seem quite long. The authors' experience shows that there is little reluctance on the part of panelists to respond to questions regarding age. In the example cited here, the researchers didn't want to highlight any particular age group, such as persons 25–44. In this case, 25–44-year-olds are the people who currently comprise the majority of hot country listeners in the market under examination.

There are several age-related issues you may want to consider. According to Jane Farley Templeton in *The Focus Group*, "Adolescents are more peer-centripetal (more conforming as a group) than adults between the ages of 21 and 65." Templeton adds, "So far, people over 65—at least those in cluster communities—seem to be more peer-centered as well." In niche research, you may want smaller age groups than you are used to

Exhibit 6.1

Seaway Research

Prize Suggestions

In a radio station contest, which of the following prizes would you like to win? (*Please rank in order of importance with "1" being your most desirable prize.*)

RANK

_____ A trip to Opryland for yourself and a friend

_____ A new truck for yourself (*please specify make and model*) _____

_____ A trip to a Biloxi casino with $1,000 cash to spend

_____ A family weekend trip to the Gulf of Mexico

_____ A family weekend trip to Atlanta with Six Flags, Falcons, and white-water rafting tickets

_____ VIP concert tickets and a visit with the stars performing at the concert

_____ A new car for yourself (*please specify make and model*) _____

_____ A pair of his and hers Rolex watches

_____ A trip to Disney World for yourself and a friend

_____ $100 cash

_____ Free gasoline for your vehicle for a year

_____ A room of new furniture (*please specify which room*) _____

_____ Other (*please specify*) _____

Which group listed below includes your age?
☐ 18–24 ☐ 25–34 ☐ 35–44 ☐ 45–54 ☐ 55–64 ☐ 65–74 ☐ 75+

Please check: ☐ Male ☐ Female

---

seeing in mass-research reports. For example, groups of 25–29 instead of 25–34. These narrower demographics should help uncover new niches and one-to-one marketing opportunities.

The Need for Ongoing Research

Testing Your Station's ID

When testing your station's and your competitors' positioning statement, or ID, your moderator may choose to ask panelists to respond verbally or to write their responses on the paper provided. For example, she may tell them, "When I say these words describing a radio station, tell me the first thing that comes to mind." She may then use examples like: *DIXIE, KISSIN', BUBBA, SUNNY, SOUTH.* If your moderator asks participants to write their responses, she may then ask for a group discussion to examine individual answers and to probe for more information.

Testing Copy and Creative Ideas

Focus groups have been the bane of copy writers for decades. Twelve people will sit in a room, listen to a piece of copy, and criticize. Or, 12 people will sit in a room brainstorming ideas. Then, a client will want one of those ideas—hatched in a Friday session—developed into a full-blown commercial concept by the next Monday. Concepts are usually judged on the ability of individuals to recall the concept. Colonel Robert B. McCormick, former owner of the *Chicago Daily Tribune*, was convinced that Thomas Dewey would become president of the United States (Figure 6.1). His recall research affirmed his belief. Recall—for recall's sake—is not necessarily a reliable measure of what is actually working in your marketplace.

Panelists Are Not Copywriters. Here are three reasons why focus groups are not reliable judges of final copy:

1. The limited number of people in one or two focus groups provides an inadequate sample base.
2. Focus groups do not provide "real-world" situations. For example, we're not hearing the commercial in an automobile or on the beach.
3. Usually panelists don't have the background or expertise that professional creative people bring to the table. Panelists may feel a false sense of power when they think they have the ability to approve or kill a commercial or creative concept. This inaccurate sense of authority can lead to biased conclusions from the participants.

To avoid the pitfalls of focus group critics, creative people learn early in their careers to put something memorable in their commercials. The

FIGURE 6.1
Chicago Daily Tribune headline: "Dewey Defeats Truman."

memorabilia usually has nothing to do with selling the goods or services being advertised. In the limited time you have to make a commercial impression, the less extraneous materials included, the more effective the commercial—no matter how well the commercial tests in recall research.

Panelists Can Help

Panelists can tell you how a commercial makes them feel. They can say whether they are amused, attentive, moved to tears, or bored when they hear, see, or read a particular advertisement. They can tell you if they remember the product or business after hearing or seeing an ad. You and your client can speculate for days about whether an ad addresses what interests a customer. The bottom line, however, is what the consumer thinks. You'll never know *in advance* of running an ad what the reaction will be unless you consult the ultimate source—the customer.

Core Demo Concepts

There is a way focus groups can make positive creative contributions to a product or service being offered by your radio station. The contribution is found in the first stages of the creative process—when you are looking for the core idea that appeals to a niche group. Focus group members usually are prequalified. They generally represent attractive niche demographics. If you receive similar ideas from three or more focus group sessions, you may have a winning advertising—or even format—concept. These ideas can make your station a winner with your clients. They can also help you refine your format to meet unique market needs. We call this philosophy the *core demo concept*.

The Weather Station. For example, after hours of focus group interviews, one successful AM radio station realized that listeners needed and would readily identify with a weather station format. So, the station positioned itself as The Weather Station and focused on that single weather image; that became their core demo concept. Ratings and revenues increased. Eventually ratings decreased but revenues continued to increase. Advertisers were sold on the importance of local weather and the extremely dedicated audience that depended on this important service. They continued to buy the station. Clients' advertising buys continued to be effective for them.

This is what niching and one-to-one marketing can help you accomplish at your station. Focus group research is time consuming. Executed correctly, however, focus group research can make a substantial contribution to your bottom line and position your station and your advertisers far ahead of the competition.

Becoming a Better Client

The more you understand the mechanics of focus group research, the better client you'll become. Better clients invest less money and get better focus group results. Here are four questions that will help you develop better focus groups:

1. Specifically, what do you *want to learn* from this research?
2. What possible alternative facts *could surface* from this focus group research?
3. What *would convince* you that an alternative idea is valid?
4. What are you *going to do* with the information once you receive it?

Share your answers to these four questions with the researchers conducting your focus group sessions. You'll receive faster answers, which could result in lower costs.

What to Expect

Many firms that conduct focus groups are quite familiar with ad agencies. Look for a research group that has some media background. Ask to interview the facilitator, the reporters, and the person who will write your final report. In some cases, all three functions may be performed by the same person. That's fine, *if* the person has a good media background. Understand that the whole process will take *at least* 30 days, depending on how many different focus groups are required. Expect to provide or pay for a place to conduct interviews that *is not* connected with your radio station (see previous section, Choosing the Site).

Providing Focus Group Research for Advertisers

In this information era, companies are launching one customer quest after another searching for the Holy Grail of customer wants and needs. Business decision makers have an unquenchable thirst for information about their customers. If you decide to join these crusades as a *provider* of consumer intelligence for your advertisers, you can place your radio station a decade ahead of your competitors. You've already seen the need for new skills in every department of your radio station. The skills needed for focus group research are acquired through study and experience. Don't become overwhelmed by looking at the mountain called *Focus Group Research*. As you and your people increase your skills at performing focus group studies, your confidence and your abilities will grow—right along with your revenues. As the Nike marketing theme says, "Just Do It."

Two Important Goals

To begin your focus group pilgrimage, your radio station may offer to conduct focus group studies for companies that consider themselves too small to work with an ad agency. Regardless of the information you uncover, your clients will learn more about their customers than they would have known without the research. And, your people will learn how to make the clients' advertising more effective because, among other things, you'll learn more about purchasing decisions made by their customers. As your expertise increases, you'll develop the confidence to conduct larger and more complex focus group interviews. When you perform focus group studies for your advertisers, you'll be accomplishing two important goals: (1) You'll be providing significant information to your clients that will help them to make better business decisions; and (2) your salespeople will be mastering the necessary skills to compete successfully in your market. Research skills will be an elementary requirement of salespeople at radio stations in the future.

Ensuring Successful Research Results

The following are essential ingredients for a successful focus group interview session:

1. *Crystal Clear Goals*—You need to clearly understand the client's objectives. You may ask your advertiser to answer the preceding four "Becoming a Better Client" questions. These are the questions that you would answer if you were the client.

2. *Knowledgeable Recruiting*—Choose panelists who have had similar experiences. For example, panelists for each interview session should be customers of your client, or noncustomers. You also will want panelists to fit the targeted age range. When you and the client create a profile of the ideal panelist, make sure the profile is not so limiting that it is extremely difficult to find people to fill interview sessions. In one case, a focus group sponsor wanted the panelists to fit the following criteria: female head-of-household who has two children or more, but one of them must be in diapers; they must own a minivan that was less than three years old; they must be full-time working mothers whose daily commute was 30 minutes or more. The likelihood of finding enough qualified participants for all the interview sessions was incredibly low. Such restrictive criteria make the recruiting process more time-consuming and therefore more expensive. When you encounter a client who insists on rigid specifications, be sure to explain the added costs involved. Also, discuss the possibility that such a homogeneous group may not have as much new information to disclose as a broader-based group.

3. *Confidentiality and Active Listening*—As a moderator, you should assure panelists that their comments will remain confidential within the

company conducting the survey. Confidentiality and a relaxed moderator foster a relaxed atmosphere and encourage openness. A group that believes they are free to express their opinions, regardless of those opinions, will give you much more information. A moderator who is a good listener will keep the group on track but will have the patience to allow a discussion that, at first, seems unrelated to the topic. A good moderator will listen to a comment, paraphrase the information, ask if the paraphrased comment is accurate, and sometimes ask for elaboration of a statement.

4. *Well-Prepared*—Perhaps the best definition of a well-prepared moderator is one who follows an unstructured interview outline. In order to uncover buying motives that are deeper than *top-of-mind*, you'll need to allow the conversation to wander down paths that pertain to your client's company. Sometimes the topic of conversation may appear to be off-course, but be patient and allow for dialogue-drifting just to make sure pertinent information isn't overlooked. After you introduce the first topic to be covered, don't try to contribute to the discussion, but do attempt to clarify statements that may seem ambiguous. Ask for further details, but don't contribute your opinions or thoughts. Never underestimate consumers and think that they are not sharp. The most effective moderator is one that causes the conversation to seemingly flow effortlessly from one topic to another. This ability increases with experience.

5. *Checking for Conformity*—While group interviews are more cost-effective and save more time than interviewing people individually, there is always the possibility of group bias. Watch for conformity within the group. Individuals are sometimes motivated to give answers that conform to group opinion so that they will be accepted socially within the framework of the session.

6. *Data Analysis*—Listening to tapes after interviews are completed will always yield a few surprises. You'll be very busy during the actual session. There will be times when participants will answer or contribute opinions at the same time. It would be tough even for a superhero to remember all of the responses—especially the simultaneous reactions. Plus, human beings have a tendency to remember things as simpler than they actually were. Also, to avoid biased conclusions, don't generalize from your focus group interviews to the larger population as a whole. Remember, your participants represent a small sampling of people who have volunteered to participate in a group discussion and may be more extroverted than the typical consumer or nonconsumer. All aspects of the interview—transcribed audio- and videotapes, recall, and written comments from forms filled out by participants—combined is the best way to analyze the data.

Starting and Conducting the Interview

Greeting the participants sets the mood for the meeting. An assistant should be placed at the main entrance to the building. A name tag on the greeter is reassuring for the panelists as they arrive. As soon as panelists are seated and have been served refreshments, all members of the research team should introduce themselves. Before the group has time to finish their refreshments, or immediately after the introductions, ask the participants to write their names, their occupations, their mate's occupation, their children's ages (if pertinent), and their age range using the paper and pen you have provided for them. You also could ask panel members to list their hobbies—assure them that they may write as many as they would like to list. This information will help you cross-check your interviewee specifications and will give you more insight into each participant.

Ground Rules

Because microphones and cameras make most people nervous, point out the equipment and explain that the meeting is being taped so you can review the session to avoid remembering it inaccurately. A joke about memory should ease the tension. Explain that you realize that most people are nervous when they know they're being taped but ask them to try to ignore the equipment. Explain that you have a lot planned for them and that soon they won't have time, or any need, to remember that cameras and tape are rolling.

Remind participants that the opinions they will be sharing are important to you and that their thoughts are not insignificant or inappropriate. Encourage them to say out loud the things that come to mind. Tell the group to feel free to respond to comments from other panel members. Assure them that you want to know what they think, what they know, and how they feel. Explain that a free-form discussion is work but that, along the way, you also expect the group to have fun.

If you're using a two-way mirror, please tell your panelists about it. Inform them that someone may be behind it at some point but you hope they'll be able to ignore it. You'll appear to be more understanding if you admit that it's easier for you to ignore the mirror because you aren't facing it.

The following example script can be adjusted to fit your situation but generally works well to explain how to handle life's necessities:

> Nature calls when nature calls. The bathroom is just across the hall, so when you must leave us for a few minutes, just get up and go. We'll understand. But come back to us as soon as possible because we don't want to miss your input.

Now is a good time to ask if there are any questions. If not, or after answering any queries, begin at some point in the room and ask a participant to share what he or she wrote on the paper. As you move around the room, try to make a comment to each person after he or she has shared information. This breaks the ice and is a good way to put everyone on equal footing.

Sample Focus Group Outline

The questionnaire in Exhibit 6.2 (the moderator's copy) is designed for a department store. *Winthrop's* is about to introduce a new line of clothing. Ownership wants to know how its target customers (and in other sessions, noncustomers) will respond to it. *Winthrop's* owners also want to know if their perspective of how they've positioned their store in the market is shared by consumers.

Additional Materials

You may want to examine radio, print, or television ads of all three department stores. If you have generic ads for a department store you may decide to peruse these. Ask panelists, "What kind of department store would run an ad like this? Are they an upscale store? A discount store?"

Your client and you may also decide to test positioning statements. For example, you may tell panel members "When I say these words describing a department store, tell me the first thing that comes to mind." Introducing new positioning statements never used in the market, interspersed with actual statements used by all three department stores, will give you insights about your client's competitors. You may decide to instruct panelists to write their reactions on paper without discussing them with anyone else. Then, you could discuss panel members' responses as a group. You'll be more likely to uncover each panelist's true perceptions rather than a *reported* view that was influenced by an outgoing panel member.

Product or Business?

Until recently, focus groups have primarily studied a product or product line. Most radio stations, however, receive the majority of their advertising revenue from direct retail. And, direct retail still represents the greatest growth area for radio. Offering focus group research to your local direct retail clients will help them see their store from their customers' perspective. They'll have a giant head start on their competitors and they will have a much clearer blueprint for their advertising copy. Every business has an *image* and a *soul*. When you help your advertisers crystalize their unique image and show their companies' soul, you will have become a true marketing partner with a rock-solid foundation for future business.

The Panelists Summarize

Just before you conclude the session, ask the participants, as a group, to give you a summary of the meeting. To encourage the group to elaborate,

Exhibit 6.2

Focus Group Outline

(Moderator's Copy)

1. How often do you shop at department stores?
2. Under what circumstances do you shop at a department store (versus a specialty store)?
3. Where do department stores fit into your pattern of buying? Do you go to department stores first when deciding to make a purchase? If searching for a particular item? If you're just looking and want to see what is new?
4. How do you feel about department stores?
   a. What do you perceive as the dominant value of a department store?
   b. What are the major disadvantages of department stores?
5. Who makes buying decisions at a department store?
   a. Does anyone influence the buying decisions?
6. Describe your ideal department store.
   a. What does it carry?
   b. What are the salespeople like?
   c. Which days of the week is it open?
   d. What hours is it open?
   e. What kinds of merchandise does it carry?
   f. What kinds of services does it offer?
   g. Is a department store a good place to shop for school uniforms? Why?
7. Compare panelists' ideal department store to three area department stores:
   a. Smith's
   b. Michael's
   c. Winthrop's
8. Do panelists see _____ department store as cautious and conservative or as cutting-edge and contemporary? (*Discuss each department store in order.*)
9. Does _____ department store feel warm or cool to you?
10. If _____ department store was a person, what would she be like? Why?
11. If _____ department store was a person, would she be like any of you?
12. How does _____ department store communicate with consumers?
13. How does _____'s communication rank with other stores?
14. Which stores?
15. Which department store is the leader? Why?
16. If _____ department store wanted to improve, how could they do that?

(Questions 7 through 13 were designed to determine how perceptions or misperceptions were formed.)

you should ask them to describe what was discussed during the meeting and what conclusions were reached. This exercise helps your client as well as the panelists to reach a point of closure and sometimes it may clarify points that remain ambiguous.

After the Interview

Immediately after the interview has concluded, most clients will want to have a short meeting to discuss what just happened. Although it's understandable that your client is chomping-at-the-bit to get your interpretation and to deliberate what they think they saw, it is premature to have this kind of review. At the end of the interview session, nobody can remember enough to analyze and interpret the conclusions accurately. Even though you may have a very anxious client on your hands, it's best to explain that a postinterview meeting could lead to some misinterpretations. This is a good time to confirm that the prearranged, follow-up presentation will be ready soon.

Analyzing the Data

Be prepared to be patient. Transcribing audio- and/or videotapes is tedious—but absolutely necessary—work. It's also the easiest part of the job. Reviewing what seems like a mountain of data before it is organized into logical sections may seem overwhelming, but don't despair. Using the suggested outline that follows, you can prepare some of your report *before* your focus group sessions. Beginning the report ahead of time will give you a sense of direction once you've collected and processed all the interview data.

Writing the Report

The following outline is not written in stone, but it will give you a sense of direction for your report writing until you develop your own system.

I. **Table of Contents**: Many novice report writers begin to hyperventilate when they realize that they'll have to produce a table of contents. Relax. It's merely a list of what is in the report and where to find it by page number. In addition to assisting the reader, the table of contents makes your report look more professional.

II. **Introduction:** The introduction describes the reason for the research; then, it restates the client's goals. The methods for achieving the goals—how you did what you did—are delineated here.

III. **Findings and Recommendations:** This section portrays some of the information you gleaned from the panelists. You do not send the transcripts to the client. The transcripts exist to help you discern the themes or patterns that emerged as a result of the interviews. When you give a volume of data to a client, your client may believe that it is the company's responsibility to review all the data and reach independent conclusions. This, of course, is not accurate. Indeed, why would clients pay you for a study if they must invest their time re-*covering* your work?

IV. **Panelists' Reactions:** After you report panelists' reactions to competitors' ads, brochures, generic commercials, and any other creative advertising materials that they were asked to examine, you may want to include some of these materials in an appendix. Many of the tangible materials probably have been given to you by the client and including some of them in the report may aid the reader in understanding panelists' reactions.

V. **Appendix:** The appendix is a convenient place to include the Focus Group Outline you used for the interview. This outline will help clarify why certain topics were discussed and why they were discussed in a particular order. In addition to the client, several people at the client's business may be asked to peruse your report. When deciding what to incorporate into the final Focus Group Report, consider what will be most useful in comprehending the conclusions.

Conflict of Interest?

Once you have conducted focus group research, examined the resultant data, and reported your findings, you'll realize that there is no conflict of interest between your station and the other media choices in your market. The information you "harvest" from panelists won't be about which radio station they listen to the most, if they read the newspaper, or which television programs they watch. As you can see from the questionnaire in Exhibit 6.2, there really is no reason to ask such questions because business decision makers want to know *why* consumers do or do not shop their stores. The advertising media delivers messages about a business to the public. Your medium may tell how a company fills its customers' needs but that is not *the media* filling the customers' needs. Media is the messenger. If you want to know which medium will be the most effective for a business, study the habits of that businesses' customers. There can be no conflict of interest when you are clearly reporting how people *feel* about a company.

A Moment of Clarity

Perhaps the most exciting thing that happens when you conduct focus group research is the moment of clarity that inevitably surfaces in a final report. You will listen to hours of audiotapes. Pages of transcripts will be reviewed. Videotapes (if used) will be watched and interpreted. The central idea, or truth, is apparent once it appears. Like any good idea, *the truth* appears obvious and simple; often, you'll wonder why *you* didn't see it sooner. You'll be anxious to share the newfound intelligence with your client. A whole new panorama of ideas will open up to you. Strategies for subsequent advertising campaigns will spill out of your consciousness as you view your client's business through a new marketing perspective.

Power

Imagine how you would feel with all the information a focus group study could uncover about your radio station, its listeners, and your competitors in your hands. Think about the business and marketing advantage you would possess. Picture how your radio station could profit from such knowledge. It's a wonderful feeling, isn't it? You can feel the excitement grow as you mentally walk through your station and envision how this information advantage can benefit every department. Now, put your advertising clients in that scenario, but instead of your radio station, picture them walking through their business. The same feelings of elation and power are coursing through them. They possess superior intelligence about a marketplace of 6 to 12 months in the future. You and your radio station have provided this information. Wouldn't you rather be the station in your market conducting focus group research for your clients than the radio stations that are competing with you?

7
Listening for Niches

Simplifying Life

You've just purchased your dream home. It's surrounded by an enormous amount of dirt and straw—your lawn. You go to a local greenhouse. They tell you about the plants in their nursery. You're told how much and how often the plants need to be watered and the amount of sunlight each type of plant needs. You hear suggestions for fertilizers with different potassium and nitrogen combinations until your head spins. Finally, they quote prices. You mumble something about talking it over with your mate as you walk away immersed in a thick fog of plants, feeds, and figures and statistics. You know you have to make important decisions, which will have an impact on the value of your most expensive possession—your home. It seems too overwhelming. All you wanted was to make a couple of simple purchases, go home, dig a couple of holes, and your lawn problems would be solved. Instead, you're so confused that you're frozen with inaction because you're afraid of making the wrong decision.

Floor Planning

Imagine how much more business that greenhouse would get if they would offer to study a "floor plan" of your lawn and then make suggestions for a layout. They're the experts. Most greenhouse operators know where certain plants will flourish and where others will languish. They understand about proportion and balance so that your hard-earned lawn can be the envy of all who survey it. You, the homeowner, wouldn't have to go anywhere else. Life could be so much simpler—at least your landscaping life would be simplified.

Looking Outside the Radio Box

You've probably heard that radio salespeople are the most aggressive and the most knowledgeable of all media reps. Unfortunately, many radio salespeople have really been like the clerks at the nursery. They've understood the ins-and-outs of their product but they haven't looked outside their "radio box" to study how their product fits into the landscape of the retailer's total picture. When salespeople concentrate on an advertiser's problems and challenges, they'll sell a lot more advertising and, maybe more important, it won't *feel* like selling. It will seem like the sales rep is just relating logical information and, then, discussing the best time to implement the plan.

Good News

Today, you know your markets are splintering. The market is *not* one big melting pot anymore. Markets are changing—rapidly; they're fragmenting more every day. To respond to fragmenting markets, radio programmers have developed more-targeted and more-niched formats. In doing that, however, it's been tougher for sales reps to use the numbers-selling methods employed in the past. As you may know, it hasn't been as effective to say, "We're number one with adults 25–54." Undoubtedly, it's going to become even less effective in the future because you're going to have to deal with formats that appeal to even more specific audiences. It will no longer be possible to say you're number one with any large group of listeners. The good news is, that's OK.

A New Thirst

Gary Fries, president of RAB, speaks with conviction about how our newest salespeople have a new thirst for knowledge. Fries notes that some stations are responding by sending some of their salespeople back to college. In fact, some managers are offering to pay for the education. Fries adds that knowledge may become the prime tool for the creation of twenty-first century wealth.

The Early Days and Today's Salespeople

In the early days of commercial radio, sales reps could survive and earn a decent income by being, basically, order takers. Many markets had only one radio station and even large markets had only a few stations. When television came along, radio reps had more competition for advertising dollars. Radio salespeople found that they needed more skills just to maintain their level of selling. Unfortunately, most radio sales reps have remained as Level One salespeople. If you go on a sales call with a Level One salesperson, you notice that they concentrate intently on selling their inventory. They say to themselves, "This is what I need to sell. Did I tell all the reasons to buy my package-du-jour? Did I use the right closing lines?" Level One salespeople can be found at most radio stations—if they have inexperienced sales reps.

Level One Salespeople—Account Executives

In the manual, *Recruiting, Interviewing, Hiring, and Developing SUPERIOR SALESPEOPLE* (1994), we identify three levels of expertise. Level One salespeople are entry-level-skilled people who present the package-du-jour by knocking on many, many doors and facing a lot of rejection, while experiencing a very low rate of closure. The rejection they encounter creates demoralized people who burn out all too quickly. Entry-level salespeople have a hard time relating to the "business generals" who run the small businesses on which they call. Level One salespeople desperately need training to survive in the era of niche marketing.

Level Two Salespeople—Survivors

These persistent sales reps have earned the badge *survivor*. These salespeople have learned the basic language of radio sales. They're comfortable with audience numbers from syndicated rating services like the BBM and Arbitron. Survivors have learned how to negotiate with an advertising buyer. They're proficient with fax-machine technology; they send one- or two-page presentations by fax. Then, Level Two salespeople wait to negotiate rates—usually by phone. By industry definition, they *are* salespeople; but, in fact, their greatest skills usually are negotiating skills. "There's nothing wrong with developing good negotiating skills," you say. We agree; but, what will take negotiators to the next level? Settling for $300 buys instead of $3,000 buys will not boost revenues to the levels wanted, and needed.

Level Three Salespeople—Multimedia Consultants

By definition, marketing consultants are in the business of . . . marketing. They get to know their prospects by using needs analyses like the Niche Marketing Analysis: Parts I and II forms. Consultants strive to create a world in which their marketing and media knowledge brings clients to them and their stations. Today, it's popular for salespeople to call themselves marketing consultants regardless of their level of knowledge. They tell their clients that a real marketing partnership will be formed when the advertiser uses their station. What that usually means is "We'll give away some prizes from our prize closet if you'll buy a remote" promises from the ad rep. If the remote isn't successful, the station will follow up with a second remote for half price or, if necessary, for free. "After all, we're in this together," the salesperson tells the client.

The Cost of Business

Typically, it costs six times as much to get a single new advertiser as it does to keep one. Because most businesses lose approximately 25% of their customers annually, a lot of time is invested uncovering new advertisers. To help ensure that the clients you have now remain your clients year after year, you'll need to invest in them by continually helping them keep abreast of *their* customers' needs and wants. The marketing techniques you're about to examine can transform sales reps into true niche marketing consultants. These are people who focus more on their advertisers and their advertisers' customers than on themselves or their radio stations' inventory.

One of the most effective questions salespeople can ask themselves may be "What can I do for retailers that retailers can't do for themselves?" When salespeople consistently do that, they're going to help their advertisers and themselves become more successful. True niche marketing consultants will also stand out from the never-ending line of salespeople trying to sell *their* products. One method to help sales reps focus more on an advertiser's customers is to conduct a survey using the Niche Marketing Analysis: Part II (NMA: II) form in Exhibit 7.1, or your own version of it.

A New Approach for a New Era

As a media rep, try this approach. Ask questions of your advertiser's customers and your advertiser's distributors. Then, suggest that your advertiser use research to verify the observations—not the other way around. Most advertisers will not question observations—if you do a thorough job. This means that you participate on your client's marketing team as an advisor *before* advertising decisions are finalized. As a sales rep, ask yourself this question: "When was the last time I sat and talked with my advertisers' customers?" By *talking* we mean *listening* to them. Unfortunately, too many media people have puzzled looks on their faces, glancing at their watches instead of listening.

Probing Questions

You've used the Niche Marketing Analysis: Part I to help you understand some of your advertiser's marketing goals. You've made a presentation, and the advertiser has said "Yes." You now have a new client. In the new era of niche marketing, however, that's not the end of the interviewing process. The next step is to use a questionnaire that focuses exclusively on your client's customers. This questionnaire is called a Niche Marketing Analysis: Part II (NMA: II). As a niche marketing media rep, you can conduct this interview with your clients to help develop more effective advertising.

Understanding customers from your advertisers' perspective is the precursor to one-on-one marketing directly *from* the customer. You *do not* have to have all the answers to help your clients develop effective advertising campaigns. You, however, *do* need some probing questions. Use the

Exhibit 7.1

Niche Marketing Analysis: Part II

1. Describe your heavy users:
   ☐ Male ☐ Female ☐ Couples ☐ Teens ☐ Seniors
2. What motivates your heavy users?_____
   _____
   _____

3. When customers buy your products (services), what problem are they trying to solve?_____
   _____

   What need are they trying to fill?_____
   _____

4. Why is that important to your customers?_____
   _____
   _____

5. Some customers are worth more than others. Describe your customers who represent the most profit. _____
   _____
   _____

6. What questions do your customers ask about your products (services)?
   _____
   _____

7. Which customers can help you get more business by referring others to you?_____
   _____

8. Which customers will never purchase your products (services)?____
   _____

   Why?_____
   _____

9. Which people would you like to become your customers?_____
   _____

10. Hearing complaints from customers is a good sign that they're interested in your business.
    Which customer complaints do you hear most often?_____
    _____

11. Who makes the purchasing decisions for your products (services)? _____

12. Are there influencers for buying your products (services)? _____

13. Who are the influencers for purchases? _____

14. Which customers are increasing their purchases? _____

15. Which customers are spending less? _____

16. Are some of your customers more concerned about price than others?
    Describe them: _____
    Most price-sensitive: _____

    Least price-sensitive: _____

17. Do your customers have the same frequency of purchase? _____
    Please explain: _____

18. Has your company's name or identity changed within the last few years?
    ☐ Yes ☐ No
    If yes, how has it changed? _____

19. Has your competition changed within the last several years?
    ☐ Yes ☐ No
    If yes, how has it changed? _____

20. How do your competitors position themselves? _____

21. How successful are they? _____

sample questionnaire to get started. As you conduct niche marketing interviews, you may develop more questions to add to the list in Exhibit 7.1.

What Motivates Consumers?

During the NMA: II interview, an advertiser may describe all the services her company performs for customers. Instead of just listing those services on your form, ask the advertiser *why* people use each service. Using Niche Marketing Analysis: Part II, you'll be able to get a much clearer picture of what motivates your client's consumers. This information will help you work with your client to create more focused ads and advertising campaigns.

Look for Commonality

Studying the NMA: II answers will help you draw a profile of your advertiser's heavy users. A vision of how these customers see your client's business will emerge. You'll begin to see her company through the eyes of her buyers. How does her store (office) fit into their lives? Think about people in the target age range. What predictable events (e.g., marriage, parenthood) may be happening in their lives? Read about niche copywriting in Chapter 15. Learn about Maslow's Hierarchy of Needs. What needs are this advertiser's customers trying to fill? Look at the suggested advertising words that appeal to people trying to fill that need. What did customers say they *love* about this store? What is the *one* thing you want consumers to *feel* or *think of* as a result of hearing your client's ads? In one sentence, write the key benefit (to heavy users) of shopping in your client's store.

Now, can you edit the sentence to four or five words? Based on the information you've collected, do you feel that your advertiser's customers can identify with those words? Will other people, trying to fill the same need, also relate to those words? If your answer is yes, you may have a very effective *theme* for your client that can be included in all of her advertising, in *all media*. When that theme becomes synonymous with your client's business—you've *branded* your "product."

A New Era in Profit Making

In most radio-selling situations, an advertiser's product is his store or office—that is, the products or services the clients sell. Sometimes, however, you may work with the manufacturer or distributor of one specific product. The process for finding a marketing theme, and finding what need the product fills in consumers' lives, is parallel to marketing a business. The cost of marketing a *product*, however, is often 2 to 8 times the cost of manufacturing that product. When selling costs can be reduced, the rewards can be great. The value of predetermining sales results should be understood by every person who has profits at stake in business, including radio stations' sales reps.

Sampling Techniques Using Direct-Response Radio Advertising. Can an advertiser tell you how much it costs to create a new user of a product or service by using sampling techniques—offering samples of a product? Ask a successful businessowner for accurate costs on each operation of her business. She can readily tell you. Ask why she uses certain processes instead of others; generally, the client can tell you exactly why the ones she uses are better. She has usually investigated competitive processes thoroughly, tested them, and knows their advantages and disadvantages. Now, ask the same businessowner for related facts on advertising and marketing. Can she tell you how much it costs to create a new customer? Can she tell you which form of sampling will produce steady customers at a lower cost? For example, will 1,000 samples sent door-to-door or 1,000 samples sent to inquiries from direct-response radio advertising produce better results?

Efficient Appeals. Ask this same businessperson if she knows the best of five or more basic advertising appeals that could be made for her products (or her company)? Ask if she has tested their pulling power. Ask if she has obtained a dollars-and-cents comparison that shows the exact cost of producing business with each appeal. If her advertising has no basic theme appeal, you can be sure the client's advertising efforts are a gamble. Guesswork is her guide. When you study the Niche Marketing Analyses for your client, you may be close to an effective theme for her business. Interviewing her customers will shed more light on the best pathway for this advertiser. A simple theme for an advertiser's business that appears to fill a need for the company's heavy users may *double* the efficiency of her advertising. Unearthing this 3- to 5-word phrase that *brands* a product (or business) in consumers' minds is the essence of any advertising strategy.

Ask Buyers

Now that you've learned what influences customers from your client's perspective, it's time to probe for more information. To find out what *truly* motivates buyers, you need to consult the supreme source—the buyer. As a part of your marketing campaign, you should suggest a study of your advertiser's customers and browsers.

Designing Survey Forms

You could design a survey form/entry blank for each client that participates in a promotional giveaway using your radio station, or you can simply design a customized survey form as a part of your service for 26- or 52-week advertisers. The forms can be used over a one-year period or for shorter periods and generally are tabulated on a weekly or monthly basis. Depending on your client's traffic counts, tabulation may be time-consuming. Remember, you'll choose the clients with whom you'll work. You'll need to decide who has the ability to give you a good return on the time you spend researching. These research services, however, will yield much higher investments from your clients. The form shown in Exhibit 7.2 is an example of a survey/contest entry form. The questions may be used, however, for intercept surveys, parking lot, or direct-mail interviews.

Adapting for the Telephone

The furniture store questionnaire can be used in its present form as an entry blank for a contest. Simply remove the two top lines to use it for an in-store opinion survey. Or, you could eliminate the first two questions and use the questionnaire for a telephone survey. You'll need to adjust the third question and eliminate the name, address, and phone number section. You may be making random phone calls or you may be using a database list (hopefully one from your own radio station). Instead of asking if someone "came into the store to find" as stated in what will now be your first question, you can ask if the person "is interested in purchasing any of the following items in the next six months." To avoid bias, eliminate the questions that ask what a respondent likes about your advertiser's store and what the owner can do to improve the store. In order to ask those two questions, you'd have to disclose your client's name, which would increase your chances of receiving answers that the respondent believes you may want to hear. Such information is not useful to your client or to you.

Geographic Targeting

You may choose to call only consumers who live in a specific area of your market. In this case, you won't need to ask your respondents' address or phone number. You already possess this information. As you can see, a telephone survey will give you much more generalized intelligence than asking questions of people (consumers) who actually visit your client's store. Yes, there is the chance that people will still give biased answers about what they like about the store and what improvements they would like to see. A minority of people may believe that positive feedback will increase their chances of winning a prize. Despite this fact, the majority of respondents will yield valuable, usable data.

Is This Necessary?

At first glance, these procedures and the time it takes to perform them may seem overwhelming and unnecessary. Experience has shown that lifestyle advertising (often called the "soft sell") is more effective and

Exhibit 7.2

Furniture Store Survey/Entry Form

REGISTRATION *(Insert Your Logo)*

PLEASE COMPLETE THIS FORM AND DEPOSIT HERE. ENTER AS OFTEN AS YOU VISIT THIS STORE. YOU MUST BE 18 TO ENTER.

Which media brought you into our store? ☐ Radio ☐ Newspaper ☐ TV ☐ Phone book ☐ Been in before

☐ Referral from: _____

This is my ☐ 1st ☐ 2nd ☐ 3rd visit ☐ Too many visits to count

I came into the store to find furniture for: ☐ Living room ☐ Bedroom ☐ Dining room ☐ Occasional tables ☐ Entertainment center/TV stand ☐ Lamps ☐ Accessories ☐ What's new ☐ What you have in store

What kind of environment do you want to create in your home?
☐ Very casual ☐ Informal ☐ Formal ☐ A mixture

If you're shopping for just 1 or 2 pieces of furniture, what are they? _____
_____

Why are you considering the purchase of new home furnishings?
☐ Refurnishing a room ☐ Remodeling a home/condo
☐ Bought or building a new home ☐ An addition to the family
☐ Need to replace worn-out furniture ☐ Other: _____

What are your 3 favorite colors? 1. _____ 2. _____ 3. _____
What are your 3 least favorite colors? 1. _____ 2. _____ 3. _____

Which room(s) are you planning to furnish and/or decorate? ☐ Living room ☐ Dining room ☐ Bedroom ☐ Family room ☐ Child's room ☐ Sunroom, screened porch, patio ☐ Home office

Please rate (use 1 for highest) in order of importance which factors determine where you purchase furniture: ___ Selection ___ Designer knowledge ___ Price ___ Financing ___ Delivery service ___ Store hours ___ Store atmosphere ___ Service after sale ___ Other: _____

What do you like about this store? _____
_____
_____

What can we do to improve our store for you? _____
_____

Age range: ☐ 18–24 ☐ 25–34 ☐ 35–44 ☐ 45–54 ☐ 55–64 ☐ 65–74 ☐ 75+

Name: _____ Phone: _____
Street address: _____
City: _____ State: _____ Zip: _____

more efficient than product (often called feature-selling or the "hard sell") advertising. Have you ever lost a client because they told you that the advertising didn't work? Too many radio sales reps have been in that position. If you hear it often enough, you'll begin to doubt your station's ability to reach consumers. You'll soon start to hear thoughts surfacing from your subconscious that ask "Is *anybody* listening to my radio station?" This scenario is one of the greatest contributors to burnout in the radio industry. Invest in your clients. Invest in yourself. Use the marketing techniques in this chapter to help you achieve success.

PART THREE
People

8

Niche Marketing Managers— A New Breed

Manager Number Five

The digital screen on the alarm clock displays 6:00 A.M. as your radio station begins to give you the first news of the day. It's Monday morning—again. You start thinking about all the things you have to accomplish at work today. Your body feels like lead. You don't know how you're going to force yourself to get out of bed and face another Monday. You vow to yourself that you're going to get off this ever-increasing, out-of-control, merry-go-round. Suddenly, you remember that you have a new manager at the station. You've seen four managers come and go in the last five years, and it seems like they all say pretty much the same thing. They all begin by telling you how things are going to be different—this time. As the weeks pass, however, their elevated plane of enthusiasm seems to descend into a fog-laden valley of apathy as each manager's words begin to sound just like the manager that was there before. Yet, even though this manager hasn't been on board very long, you actually sense a difference.

Manager Five—a moniker the salespeople have bestowed when out of earshot—doesn't ask you to make-as-many-calls-as-possible today. Instead, Manager Five (MF) wants you to focus on the *quality* of each call. You don't have a deadline for getting out of the station each morning. In fact, if you need to spend an *entire day* at the station conducting research and creating marketing presentations for your clients and prospects, that's OK with Manager Five! All you have to do is share your work with MF, discuss some of your strategies, and present your advertising plans. Recently, you've noticed that you're feeling a lot better about the sales calls you're making.

The advertisers seem to be happier to see you because you're not stopping by to see if they have "anything planned." Your calls are focused on your advertiser's customers now that the two of you are always looking for ways to find out more about those customers' needs. You use that information as a guide for more advertising. Now that you think about it, you realize your closing ratio has also increased.

You glance at the clock radio and notice that it's now 6:05. It's time to get up and mentally prepare for that big presentation you're giving this morning at 9:30. You're really excited about the information your research uncovered for your newest client. You're convinced they'll love it. You know, this manager just might be different. But you don't have time to think about that now, because it's going to be a *great* day!

What Niche Managers Need to Know

The 9:30–4:30 Rule

The preceding scenario could be one of your sales reps describing her job at your radio station after you embrace niche marketing techniques. In order to bring salespeople into the niche marketing future, however, management will need to gain a new perspective on the sales department. As salespeople begin to conduct more Niche Marketing Analyses, they'll need to spend more time at the radio station conducting research and preparing marketing presentations. Unfortunately, many stations' sales reps are told to be out of the station by 9:30 A.M. and not to come back before 4:30 P.M. Some managers actually believe that if salespeople aren't at the station, they must be out selling. The 9:30–4:30 rule represents the make-as-many-calls-as-you-can thinking of the past. It'll be different in the future—at least at stations that are going to offer state-of-the-art marketing services. Sales reps will be spending at least 40% of their time at the station conducting research, preparing presentations, and studying marketing strategies for their clients because those activities will help yield the most sales.

Fiction and Wish Lists

Not only will management see salespeople differently, their relationships will change. Goal-setting will no longer be a top-down process. Managers already know it doesn't work. Middle managers are told how much billing the station needs for the next year—or quarter. In turn, sales managers tell salespeople how much billing is needed. Reports are required from each sales rep to show how billing goals will be reached. Salespeople spend a lot of time writing some of the best fiction and wish lists ever devised. It's not that they don't want to accomplish these goals—they often don't know how, using the limited resources available to them. When salespeople begin to see themselves as business partners with management, they'll start to believe that they really do have some control over their professional lives. Optimism will grow. Cooperation will multiply. Sales will increase. Turnover will decrease. Your radio station will become more stable financially.

Bye–Bye Prima Donnas

As salespeople focus more on their advertisers' customers and their advertisers' marketing challenges and less on their need to sell a specific amount of advertising time, you'll find that your sales department will become more customer-driven. More emphasis will be put on service than ever before. A service-oriented sales staff will find that they get more repeat business. To avoid being left in the dust of your competition, every department at your radio station needs to be customer-driven so that decisions affecting clients can be addressed fast. Your sales staff needs to understand how to be flexible without giving away the station. Advertisers are impressed when sales reps can make on-the-spot decisions. Naturally, the radio station should be able to remain cost-effective after sales reps' decisions are implemented.

Shrinkage

Markets are splintering, ad budgets are being down-sized, and media choices are multiplying faster than rabbits can procreate. Consequently, radio stations are feeling more and more pressure to justify their rates to advertisers based on perceptions of audience size and how far the competition is willing to cut *their* rates to take the buy away from you. Many sales managers give their salespeople the authority to cut rates when they're with a client, if it means they'll get the buy. Unfortunately, salespeople who have very few negotiating skills and even less marketing knowledge, often will settle in at the bottom of the rate card and make all their deals *from* there. With the bottom of the card as the starting position, you're all too familiar with the ultimate downhill destination. Your profit margins continue to shrink.

A New Negotiating Tool

In the new era of niche marketing managers, you'll want your salespeople to have the flexibility to negotiate a sale but you'll give them a distinct negotiating tool—extra services—which will be built in with the advertising proposal. Your salespeople can act fast and be flexible. They won't be running to the bottom of the rate card or giving a free remote just to get the advertiser to sign on the dotted line. You can expect your profit margins to stabilize and even increase.

Unique Insights

Salespeople, some of the frontline employees of a radio station, have a perspective on your market and your advertisers that management can never have because managers aren't seeing their clients face-to-face and getting feedback on a regular basis. When sales reps believe that their sales manager values their input, and they trust that it won't be used against them, managers will be amazed at all the creative ideas they'll hear. Salespeople, as well as other departments' employees who deal directly with advertisers, can offer a wealth of information about improving billing, sales offers, production, and just about any other function at your station.

Who Has Power?

Some managers believe that having power is being able to tell the people you supervise what they should be doing and then checking on them to see that they're doing it. These managers feel full of power because they give other people orders. But, who really has power? A seasoned salesperson once worked at a radio station where each sales rep was within about $1,500 of their monthly billing goal but collectively the sales manager was $10,000 short for the month. This manager continually told the salespeople that she was in control and that someday they might get a chance to be in a position of power like hers. The sales staff was convinced that she didn't care about them and, as a result, they felt they didn't need to help her. After all, the salespeople reasoned, no one was going to fire the entire sales staff. Unfortunately, this situation yielded a lose–lose outcome because the sales manager didn't realize that everyone has power.

Unlimited Power

If a sales manager believes that giving away power diminishes one, the manager also believes that power is limited. These managers fear that if they give some away, they will have less. Actually, just the opposite is true. Power is unlimited—the more you give away, the more you will have. Give some away. When you empower your salespeople to make decisions for themselves, they'll begin to make *better* decisions. People who know how to make good decisions increase their skill levels. As salespeople gather more skills, they will gain more self-knowledge. They will like themselves more and be open to the possibilities of *more* mental expansion. Will your power decrease because their personal power has increased? On the contrary, this may be the only time you will have real power.

A New Breed

At a recent marketing leadership conference in Dallas, Gary Fries, president of the Radio Advertising Bureau (RAB), talked about the new breed of radio sales managers coming into the radio business. Fries noted (personal interview) that he sees

> ... a different person now. The sales management people of the 1980s are filtering out of the business. Those were people who grew from salespeople to be sales managers by virtue of being an outstanding salesperson. When they

were appointed sales manager, their number-one achievement was ego driven. They were the boss. They were the sales manager and then—later on—they often became the general manager. Today, managers are students. These are people who have entered the arena. I see a change. They have a thirst for knowledge. If you can give them the opportunity to learn, they are willing to apply themselves and take the knowledge offered.

Gary Fries also points out the need for management and cross-media training. The *challenge* for successful managers is to have salespeople know more about competitive media than the media salespeople with whom they compete. Niche marketing managers who share this knowledge can transform salespeople into true niche marketing consultants.

Co-owners

Okay, so you're convinced that there might be something to this new way of working with salespeople. Where do you start? Your first goal may be to get your salespeople to think of themselves as co-owners of the radio station. Indeed, every employee of your radio station already owns a *mental* piece of it. Until now, however, the ownership idea may not have been openly recognized and encouraged.

What's So Great About Ownership? Ahh, nothing feels quite like being "the boss" when you're walking through your radio station. Yes, you get a lot of rewards and freedoms but, after all, you take the risks. The loan at the bank is in your name. If you default on the loan, it's you who pays the consequences—or is it? If you were no longer at the station, would other jobs really be affected? Yes. Every employee at your radio station has a vested interest in the continued success of the company. Jobs are on the line every time there is significant change within a business. When your receptionist answers the telephone and makes contact with an advertiser, or a potential advertiser, your company and its profits are on the line.

Every time a salesperson makes an omnipotent sales call, that decision affects the well-being of your company. That is, they drive by a business and take "No" from the road—without leaving the car, salespeople often decide that the potential advertiser couldn't possibly want to buy from them right now. Additionally, when salespeople accept low rates for an advertising buy, there is an impact on every employee at the station from the potential lost for reinvestment in the radio station. You won't be able to purchase that new piece of equipment for remotes that would help your station stand out in the market. You won't be able to send people to a seminar that promises to teach new techniques to increase revenues.

Things Would Be Different If . . . Things would be different if you, the owner, were the person driving by that business. You'd think of all the potential for gain at your radio station and you'd welcome the opportunity to make the next sale. You'd hang in there during the negotiations, remembering why the client should pay your rates even though they're higher than the radio stations the advertiser usually buys. Yes, things sure would be different if you were the on-the-street rep. Then, you remember—you can't spend all of your time calling on clients. You have many other responsibilities and duties to perform at the station. One person couldn't possibly sell and service enough clients to amass the billing your station needs. If you could only clone yourself! Well, there is a way.

Becoming Business Partners

1. Begin by telling salespeople how essential they are to the success of the radio station. Each of their personalities leads them to uncover and sell advertisers who feel comfortable with their unique type of sales perspective.

2. Describe how you couldn't operate it without them. Dare to confess that you couldn't operate the sales department by yourself. There aren't enough hours in a day for one person to call on all the advertisers needed by a radio station.

3. Tell them that to be the most successful, they need to think of themselves as co-owners of the radio station. In fact, they already are co-owners mentally with every other employee and you.

4. Describe how *every* employee impacts a business. Each employee's contributions are important to the success of the entire station. From the custodial staff to the general manager, each person represents a valuable link that creates the total chain.

5. Explain how each salesperson has a vested interest in every advertising sale made by *every* salesperson at the station. When the entire sales department reaches its billing goals, the company becomes more economically sound, which, in turn, gives all salespeople more job security. As salespeople take the time to help other sales reps at the station, they also will be helping themselves.

6. Tell your salespeople that in order for them to be more successful, you understand their need to have more information about sales reports. In fact, to become more effective business partners with the radio station, you realize that salespeople need to understand and have access to reports from other departments as well.

7. Detail how you would like to see their creativity unleashed. For example, instead of creating a promotion for the sales department yourself, give the project to one salesperson you feel would have a special interest in the event. Include a general guideline for creating the promo-

tion but *let* the sales rep actually design the package, including any accompanying promotional ads. The most important gift you can give the salesperson is to accept that the promotion will be different than the end result *you* envision, and that's OK. Actually, it's better than OK. You may find that the sales rep includes ideas and provisions that have never been tried at your station. Just be supportive and keep an open mind.

Ask the salesperson to present the promotion to the entire sales staff at a sales meeting. Remind him to justify the rates he is quoting—first from a radio station owner's perspective and then from the perspective of the competitive marketplace. This occasion gives you an excellent opportunity to teach your salespeople about the factors you use when determining a promotional event—whether you use a ten-times factor to price promotions or a lesser factor. Show your reps the way you determine how much a promotion will cost the radio station and, then, how to multiply that cost by the particular factor your station uses. Explain how a percentage of other station employees' salaries, costs of maintaining equipment, expenses for health plans, and other costs of conducting business need to be considered when pricing an advertising offer. Sales reps will gain an important insight regarding why they should quote rates at a certain level and then stick to that level. You'll also be encouraging your salespeople to utilize their creativity. The whole station will benefit.

One Manager's Self-Talk

At this point, some managers may be cringing at the preceding scenario. They may say to themselves, "I can't let my salespeople create promotions! We'd be out of business by next week! My salespeople just don't have the understanding to put together successful promotions! They just don't have what it takes!" The emotions underlying these words, however, may uncover very different reasons for your objections. "If my salespeople learn how to create really successful promotions, who will need me?" your subconscious asks. "If my sales staff doesn't need me, top management won't need me either. I'll be in a bread line by next week!" your mind shrieks. Wait. Let's move to the rational side of your brain and examine this a little closer.

In the past, many people have been promoted to the position of sales manager because they were the best sales rep on staff and management needed a sales manager. It's less expensive for ownership to hire from within because a current employee already knows the operation of the sales department along with a rudimentary knowledge of how other departments function. A current employee also understands the station's basic marketing premise. All this knowledge saves the station time and money because it hardly slows down the forward momentum of the sales department when installing a current sales rep as sales manager. OK, many sales managers are also promoted from within because they are great salespeople. But, what makes a great sales manager?

One thing that adds to a sales manager's success is knowing how to clone themselves to build an outstanding sales team. Basic survival teaches that two heads are better than one and three heads are better than two and so on. In most cases, one person cannot produce all the billing a station needs; it takes a team to produce the sales needed by your station. A team can also produce more creative promotions because more minds are involved. Your job is so much more complex than producing promotions. Your job is to help your salespeople discover the greatness within themselves. Now, take out your mental binoculars and look a little farther down your career highway. Isn't a sales manager's ability to build meteoric sales teams seen as a rare jewel in the radio industry?

Salespeople Want to Succeed

Sales managers who fail to achieve the levels of success they desire usually fail to bond with, listen to and hear, and really look at salespeople; and they fail to help salespeople imagine themselves as autonomous. A sales manager's ability to communicate determines her ability to manage salespeople. If a sales manager took a survey of the sales department, it would probably be found that every salesperson wants to succeed. But wanting to succeed, knowing how to succeed, and having a burning passion to succeed all may be ideas that live at different addresses.

Bonding. There's an old marketing adage that says "People don't want to buy ¼"-drill bits but they do want ¼" holes." If salespeople were asked if they wanted to bond with their sales manager, the questioner would probably be laughed out of the room. Ask "Do you want your sales manager to remember how incredibly tough a sales rep's life is on-the-street?" and you'll see salespeople nodding their heads up and down—without smiling. Their nods translate as "Yes, I want to be appreciated and respected for what I do." One of the first steps toward forming a bond with salespeople is to begin to treat them as business partners. It's really OK to admit to them that you want to see them win because when they are successful, you will be successful.

Build a picture of how you fit into the organization and the role you fill in their business life. Explain how you are there as a supportive team member. Yes, you are required to draw *general* guidelines for the department, but you expect, with their help, to write and rewrite the guidelines as you—and they—proceed. Partnerships begin when you respect a salesperson, admiring who they are and what they do. Each of your salespeople has a selling skill, or a people skill, you don't possess. Honor and encourage the uniqueness of each person's skills. Look for ways to praise each salesperson. Seek opportunities to speak admiringly about salespeople in front of co-workers and advertisers. Be genuinely grateful for their contributions.

Listening and Hearing. *Listening* is a learned skill. When you really listen to another person, you also participate. Establishing eye contact is important so that the other person knows you're paying attention. Paraphrasing a person's words and feeding them back in the form of a question also tells the speaker that you are indeed listening and interested. For example, "If I heard you correctly, you feel that sales meetings at 4:30 on Wednesday afternoons aren't very productive? Is that accurate?" *Hearing* is making sure the person speaking is communicating what they wish to communicate. Giving paraphrased feedback is a way of checking that you heard accurately what someone was trying to say. Human beings are some of the poorest communicators on the planet. Be patient. Your people will appreciate a sales manager who really listens. They won't forget.

Looking and Seeing. Look at your salespeople. Really *look* them in the eye and silently say to them "I *see* the humanness of you. I respect that you have fears just as I have fears. I know you've experienced rejection and that you're hungry for approval. You are a person just like me who wants to connect and to be liked. I really do want to know the person—not just the worker." When you look at your salespeople and think these thoughts, they will respond—intuitively. They will be more open to the information you want to share with them. You will build trust. Be ready to cut off your arm before you betray that trust.

A Sales Manager's Goals
1. I want you to feel good about who you are at this moment, as a person.
2. I want to help you bring out the best in you. It is so rewarding to see a human being who sees his own value. It opens up the opportunity for the person to see the value in others—including clients.
3. I want you to *love* your job as a salesperson at this radio station. My goal is to help you adore all the hundreds of things you do as a salesperson *so much* that you want to *jump* out of bed in the morning so you can begin your day! What greater gift can you give other human beings than to help them *love* their lives?

Oops!
Far too many radio station managers feel that it's their job to monitor salespeople and correct mistakes. When the manager catches someone doing something wrong, there is often a one-way discussion about never making that mistake again! Regrettably, this discussion may take place in front of co-workers, rather than in private—or at all. Salespeople frequently feel

like a child being reprimanded by a parent instead of a business partner at a radio station. The new breed of sales manager searches for salespeople doing things right. By all means, sing it from the rafters when you find salespeople doing something positive and productive. No one gets enough positive reinforcement.

Continue to expect mistakes, but use those mistakes as opportunities to discuss how your system can be improved to avoid the same results in the future. Encourage salespeople to make mistakes faster! Place signs throughout the sales department. The pictures you see every day in your home and at your workplace exert a subtle influence on you. Placing signs on the walls in the sales department that tell salespeople it's OK to make mistakes *re*-enforces your belief in them and frees them to try new things. Tell them if they're not making mistakes, they're not living. Don't be afraid that you'll be encouraging irresponsible behavior. You know that you invest a lot of time telling your team what you hold them accountable for. As people realize that making mistakes is expected and seen as an opportunity for growth, they'll feel freer to become more creative. Increased creativity gives birth to increased enthusiasm; as you know, enthusiasm is contagious. Advertisers buy more from enthusiastic salespeople.

Opening the Secret Diary—Start with the Billing Printout

Where's the billing today? It's a question frequently asked by the general manager (GM), general sales manager (GSM), and salespeople. In the past, the full picture of the monthly billing has been seen only by the managers. Often, in an attempt to motivate, salespeople were shown their billing figures and told that other sales reps were outselling them. Salespeople talk among themselves, however, and the sales manager's efforts to motivate by intimidation have usually backfired. As the new breed of sales manager in this new era of niche marketing, you'll find that when you share billing information with everyone in the sales department, as well as everyone at the station, you'll begin to get marvelous feedback about ways to improve productivity.

When all employees begin to think of themselves as business partners with the station, they'll start to see that doing the best they can in their position will benefit everyone and increase their job security. Prove to people that their input is valued and wanted. Every employee will begin to feel that they're an important part of the radio station team. When people feel they're valued, they will begin to see their work as more valuable. Their sense of pride in what they do will increase. They will look for ways to improve *their* part of the company.

Reality Is Better Than Speculation

Sharing the billing printout with all employees may seem like a revolutionary concept. A mountain of questions, asking why billing information should be kept confidential, may be running through your head. Everyone at the station already has some idea of where billing is at this moment. People confide in each other. People read memos—even if they aren't meant for them. It's a fact of life. When billing figures are good, you shout it throughout the station. When you're secretive, speculation runs rampant. Reality, with a plan for improvement, can be reassuring for all employee/business partners. It takes everyone working together to make a great radio station.

Don't Stop with the Billing Printout

Share other departments' reports with your salespeople. Wait a minute! Why would any manager want sales reps spending their time reading reports about other departments? You're looking for increased efficiency, not decreased. They need to focus on making *their* department better. It's really none of their business how other departments are progressing. Actually, it sounds like there would be a lot more confusion. Not everyone in the organization knows how to interpret the reports. You can see the chaos developing now. After all, that's why there are managers at radio stations.

> At one radio station recently, a lot of money was spent on outside programming consultants. Ownership felt the money was well spent because these programmers had given them good advice over the years. Music tests conducted at local auditoriums had yielded valuable insights about the music tastes and information needs of the station's demographic target. Salespeople weren't invited to attend these sessions, however, because it would have taken valuable time away from selling. Salespeople were also told that they might be recognized by audience members, which might sway the results. No explanation was ever given to the salespeople about the methodology for conducting these programming surveys because that was a function for the programming department. It was hinted that the program director would feel threatened if salespeople inquired too much about the way things *worked* in programming.

Unfortunately, when this station's salespeople were in front of a client, questions were asked about why the advertiser should pay a higher rate for their station when the station across town, *which played the same kind of music,* was only half the price. It was impossible to adequately explain to the advertiser how extensive research had been conducted to find out what attracted a specific group of consumers to listen to their station and of all the work involved in finding out what kept those consumers tuned in hour after hour, day after day. Oh, the sales reps could give *general* descriptions about all the testing, but since their knowledge was limited, they couldn't describe the process in detail. They couldn't take the advertiser on a mental, step-by-step journey through a session—opening the world of media programming so that the client felt like an insider at the radio station—even if the advertiser had never set foot in the station physically.

One of the first rules of marketing tells us that to be a leader in a category, we never *say* we're the leader—we show it. If we're bright human beings, we don't have to tell people—we show it in the many interactions we have with others. As you know, people buy from salespeople whom they believe understand them and their needs. They also buy from people they feel they know. The most successful salespeople make their clients feel like they have a relative in the radio business. One of the ways successful salespeople convey all these attributes is to share information with clients about the inner workings of a radio station. Understanding *more* than the basics of the programming department and relaying this information, when appropriate, is one of the keys to selling more advertising and to keeping clients longer. The programming and on-air staff could also benefit from understanding and participating in the sales procedure, including making calls with salespeople. Let the information sharing commence!

A Sense of Control

Let's go back to the first rule of the new breed of niche marketing managers. Managers need to change the way they think about managing and training salespeople. Of course, any manager can slip back into an old way of thinking from time to time. Everyone regresses somewhat when faced with new challenges—at least in the beginning. Sharing information with salespeople— actually all employees—is imperative if your station is going to survive and grow. Experience has shown that there is a lot of report-reading in the beginning, either out of curiosity or to test that management really means what they say. As employees adjust to their newfound autonomy, however, they'll begin to ask for and focus on reports that directly affect their individual departments. The end result will be that they'll have a better understanding of their place in the whole organization. Just as the Information Revolution is helping us to see ourselves in new ways, sharing information within our radio station will help all employees feel more like business partners with the company. The increased sense of control over one's life leads to more job satisfaction, which leads to more creativity, which leads to a more efficient and effective radio station.

The Big Picture

With all this talk about empowering employees, where do you, the manager, fit in? A part of your job description is monitoring salespeople—or is it? The answer is yes—but in a different way. Hopefully, you hire or inherit capable salespeople. You give them appropriate training, positive feedback, and then, get out of their way so they can do the job you want them to do—you hope. The first step toward less monitoring and more empowerment is to paint a crystal-clear picture for your salespeople, showing them what you expect from them. This simple process may seem like a contradiction. Why would you tell your salespeople what *you* want from them? Isn't the niche marketing manager supposed to be a partner with each salesperson rather than a dictator of wants?

Yes, but successful relationships, personal and professional, need guidelines. A picture of what you expect from your salespeople is not the same thing as setting goals for them. Rather, a mental picture can paint a vision of the future of your radio station. For example, your company's vision may be to become a shining example for the radio industry. Your station may aspire to give advertisers the best service they have ever received from a radio station. Your sales staff may offer the most up-to-date marketing techniques available to their clients. Your programming staff may have made a commitment to a daily dialogue with listeners in order to serve the market better than any other media.

When people understand where the station is headed and how they best fit in the picture, it becomes easier for them to see themselves as business partners. For instance, instead of telling salespeople *what* to do, show them *how* to focus on an advertiser's customers and on what motivates those customers—don't concentrate on which inventory needs to be sold today. It will be easier for your staff to understand what you expect from them and why. Salespeople perform many tasks in the course of a day. Prioritizing tasks will be easier for your sales staff when they have a clearer picture of what you expect from them as a contribution to the big picture. When people see themselves as an important part of the process, they will feel more commitment to the radio station's goals.

Personal Counselors

A picture of what you expect from salespeople needs to be accompanied by an equally distinct description of what you will give to them as their manager/helper. A sales manager who empowers salespeople asks them, "What can I do/provide for you that will help you with your clients?" If John F. Kennedy had been a sales manager, he might have said, "Ask not what your salespeople can do for you. Ask what you can do to serve, assist, contribute, and empower your salespeople." The most successful sales managers focus on how to help each salesperson develop; they concentrate on being a personal counselor. Help your salespeople talk-out a situation and help them gain insights. Self-motivated salespeople are the most productive. Be a sales manager who is warm, approachable, and nurturing. Many may disagree with this approach, especially the nurturing aspect, but try it. Look at the turnover of salespeople and sales managers. How well have others managed without nurturing?

Confess

Share your humanness with your salespeople. You, no doubt, have been on the *other* side of the sales manager's desk. It can be so daunting to be called into "the boss's" office—or just to go in for some information. Some sales managers believe that to admit they're less than perfect is the same as encouraging their salespeople to make the same gaffes. You remember the times *you've* taken *no* from the road or gone shopping in the middle of the afternoon instead of calling on advertisers. Your salespeople will have these experiences anyway; it's inevitable. Everyone needs these events to learn from them. Your acknowledgment of your actions will help your salespeople identify with you. They'll begin to see you as someone who understands them and their challenges. A salesperson is much more likely to confide frustrations to someone he believes will empathize and make helpful suggestions rather than just criticize. It is easier for salespeople to accept their own mistakes when they see you admit yours—and forgive yourself for making them. Life can be so much more fun when you laugh at your mistakes rather than hide them. Anyway, there are no mistakes, just opportunities for growth.

The Encourager

Become each salesperson's biggest fan. Sing their praises to the general manager. Many sales managers invest a lot of time singing their own praises to the GM because they're trying to protect *their* position. They feel they need to say, "Look at what a good job I'm doing. You could never do this without me." Some managers feel that if they praise a salesperson to the GM, the GM will begin to look at that individual as a possible replacement for the sales manager.

Consider this. If you dare to be the kind of sales manager who praises your salespeople to the boss, the boss may perceive you as an outstanding manager. After all, isn't your mission to build an exceptional sales staff? But what if your worst fear is realized? Your GM decides to replace you with one of your salespeople? Will your life end? Will you lose esteem in the industry? Will you lose the respect of your peers? Will you lose material possessions? Will your mate think less of you? If you take the time to sincerely answer these questions, you'll see them for what they are—fears you carry with you all of the time. But they are just that—fears. What if your worst fear is realized and you lose your position? It is *not* the end of your existence. You may, or may not, realize your fears but you will gain enormously in the long run.

Visualizing the loss of one's job as being beneficial may be difficult to identify with until you've been through the process, but it may precede another door opening for you in an organization where your skills could be more appreciated and rewarded. You may look back at your previous position and be incredibly grateful that you've moved on with your life. Whichever paths you take in life, you'll find positives and negatives along each way. Each path is right for you at that time. Take chances. Dare to be the new breed of sales manager that heralds the triumphs of each salesperson on your staff.

Lazy Salespeople

Have you ever heard a sales manager talk about having a lazy salesperson who just couldn't be motivated? The dictionary defines *lazy* as not easily aroused to activity; also a disinclination to work or to take the trouble. Most people could be described as lazy at some point in their lives under the definition of a disinclination to take the trouble. But what about the salesperson who has a disinclination to work or who is not easily aroused to activity? If you have a salesperson who fits this description—and who hasn't at one time in her career—take a closer look at the specific activities the person most often delays. At first glance you may want to answer "Everything!" When you begin to watch that person's actions, however, you may find that the procrastination is limited to specific activities. Is it the most important things the salesperson puts off, or is it something small, like finishing paperwork for internal use. Are the postponed activities in an area where a salesperson has a lot of ability or where his skill level is low? Watch for patterns of avoidance. Does one salesperson spend much more time than is reasonable putting together one wonderful presentation after another, but never gets in front of the client to share the information? What about a salesperson who closes a lot of sales but then doesn't turn in the paperwork to get the client on the air in time for the client's big sales event?

Facing Fears. A procrastinator's inactivity is usually the result of fear, not laziness. Those whose predominant thoughts focus on fear of failure are imagining that their efforts won't measure up to the expectations of the sales manager, the client, or whomever asks them to do something. The result is that they do nothing. As you might expect, just the opposite is true for someone who fears success. A salesperson may reason that if she turns in an outstanding performance either the sales manager, her mate, or both, will expect that level of commitment in the future. The result may be a less-than-stellar performance or becoming frozen with anxiety. Regardless of the fears, the results may be the same—a perfectly capable salesperson who doesn't achieve goals.

Uncovering Busywork. One way to begin helping your procrastinators free themselves from the bonds of avoidance is to ask them to list all of the activities they carry out in the course of a week. The longer the list, the easier it will be to illustrate your point. Your point will be to show which activities actually lead to sales and which are simply busywork that can be executed by members of the support staff at your station or performed later in the day when activities requiring more brain power have already been accomplished.

Reasons Why... Many procrastinators appear to be very busy people. They often overbook themselves with commitments to avoid the activities they fear. In order to help your procrastinators, track the excuses you hear from each one over the course of one or two weeks. Many people are so accustomed to hearing themselves give certain excuses that they are no longer consciously aware of what they're saying. After you've compiled your individual excuse lists, schedule one-on-one appointments with each sales rep. Of course, it goes without saying that these meetings are the most productive when you explain at the outset that you've been developing a list of *reasons why* certain tasks are not being accomplished, either in a timely fashion or at all. As long as the person understands that you're not trying to find fault but are really trying to help them, they will be more open to an honest discussion. If they really trust you, they'll actually be relieved to discuss the anxiety their procrastination has caused.

As you discuss each excuse, examine what happened right before the reason was given. Discuss why each incident prompted an excuse to put something off. Getting a procrastinator to see what they're doing is the first step to helping them help themselves. Some experts suggest continuing the delaying tactics for at least another week so the person can consciously watch for her responses. Becoming aware of the choices one makes can be the first step to an empowering experience.

A Highway of Uncertainty. Setting *realistic* goals with your procrastinator is a good way to help them get to work. Start small and caution the salesperson to be patient. Setting too many goals and expecting too much change in a short time are two of the reasons people delay in the first place. Be specific when choosing goals. A goal such as "I want to sell four new clients this month" is admirable but far too general. A general goal needs to be accompanied by details. Focus on specific steps the salesperson can take *each* day to reach a goal by the end of a month. A clear picture of where they're going will help both of you outline the steps for getting there. Sometimes, one day at a time is too long for a procrastinator. One call at a time or one decision at a time may be all they can handle in the beginning.

Follow-through from the sales manager each week may seem to defeat your purpose but you'll be helping someone develop productive habits. Successes, even small successes, are essential to show procrastinators that positive change is possible. Remember, the salesperson's legs are wobbly on this highway of uncertainty. Reward successes. A note congratulating someone, or a compliment given in the hallway, will go a long way toward giving the person an incentive to continue the new productive actions. Even if the sales rep didn't accomplish every goal, or do it in the way that was planned, celebrate achievements—it's much more important to feel good about the efforts than to give a perfect performance. If the rep feels good about what has been accomplished, she will be much more likely to repeat the efforts and then the ability to follow-through improves. People frozen from activity—afraid to take any action—often focus on what they didn't achieve rather than on what they did accomplish. Help them focus on the positives.

Inner Peace

Sales managers are asked to make decisions concerning the sales department and salespeople every day. Many people, including those outside your department, may be affected by your decisions. You know you can't make everyone happy. You want to make decisions for the good of the individual salesperson and for the good of the organization. If you must choose, put the salesperson *first* over the organization because then, and only then, will the organization survive. If you're agonizing over a decision you must make, take a moment to quietly go within. Always look at the bigger picture. Ask yourself which decision will serve the highest good of that person. You will know the answer. Then, let go. You've made the best decision for that time. Be at peace with yourself.

Managing

Managing is not about your ability to *manage* others but more about your ability to help people bring out the best in themselves. The best managers are the ones who help people develop beyond what they thought they were capable of achieving. Helping others expand can never be accomplished through intimidation, the use of threats, or ultimatums. Negative ways of dealing with various situations may change actions temporarily, but in the long run, negativity will only serve to alienate people. Positive reinforcement is the most important thing you, as a manager, can give. When people feel good about what they're doing, they're much more likely to repeat those actions. If you're good at the game of *Monopoly*, you like it; if you don't experience winning once in a while, you are likely to say you don't want to play. If you're not winning, the game can seem to go on and on; but if the tide turns and you suddenly have a chance to win, you get excited. You become wide awake. Your breath quickens. You speak louder. You're *involved*. You can *see* yourself winning. The life of salespeople is very similar. Help them see themselves winning.

The Journey

Being a great writer is not just about one's ability to write, because one never writes "the great work" on one's own. People who are great communicators through writing are able to tap into the universal stream of information that has been collecting and growing since the beginning of humankind. We (the authors) once thought that it was such a waste when an older person died because so much knowledge and wisdom was lost. We now admit that this is short-sighted thinking. You often hear of novelists proclaiming that they don't know where their stories originated—"The characters just seemed to come alive and develop the plot on their own as if they were actually living it," writers confess. While they've been tuning into this vast ocean of consciousness for inspiration, this same source is available to you as a manager who is devoted to helping people feel better about themselves, their job, and ultimately their clients. It's the *process* of focusing on bringing out the often-undiscovered greatness in salespeople that makes one a great manager. Remember, the journey to becoming an outstanding sales manager *is* the destination.

9

Niche Marketing Sales Consultants—Their Difference Is Knowledge

What Holds People Back?

The main reason people get stuck at a particular level of development is a lack of faith in themselves. They refuse to let go of the familiar in order to reach for the unknown. They clutch familiar assurances to their chest, focusing on past achievements, hoping that somehow previous risk-taking adventures will carry them through the remainder of their lives. They aren't willing to give up their positions of achiever, accumulator-of-power, or reputation—however temporarily—to reach for the next level of development and self-awareness. To reach new horizons one has to be willing to let go of familiar landscapes.

Effective Training for Niche Consultants

The First Compliment

For a radio salesperson, one of the most difficult aspects of the job may be cold calling. Add repeat-calls-on-difficult-people to that and you may have a daunting job description because those two activities could occupy two-thirds of a sales rep's time. Within a few days of starting a sales career, most people are told to give a stranger, or difficult person, a compliment to establish rapport. But make it a genuine compliment, you're told, because the person will know the difference. You're taught that the purpose of the compliment is to relax the prospect and that this somehow will help the person like you. After hearing about 10 "salesperson compliments" a day, most prospects are not only insensitive to them but downright disparaged. What most sales trainers fail to tell you is that the purpose of the compliment is—or should be—to enrich that person's life in some way—not to make him or her like you. The following may help you get beyond the "first-compliment" stage in developing a relationship with an advertiser.

Before the Call

Take a moment before you go into a client's or prospect's business and have a mental discussion with yourself. Even if you've never met the person, try to imagine the decision maker inside the building you're about to enter. Remind yourself that this is a person just like you. Advertisers have obstacles and challenges in their lives that they're trying to handle to the best of their ability. They face rejection; they want to be liked by others and to connect with others, just like you. There is no client who is that different from you. Picture yourself walking into the building. Friendly employees help you *easily* find the person you need to see. Imagine yourself feeling very comfortable in the situation as you explain who you are and why you're there. In your mind's eye, see this person being positively responsive to you and willing to listen to what you have to say. Imagine this person's eyes. Look into a client's eyes and silently say to him

> I'd really like to get to know you. I truly believe I can help you and your business become even more successful. I would like to salute your accomplishments and understand your challenges—both business and personal. I want to know your goals and I want to be a part of helping you reach them. When our meeting is finished, I want you to feel better about yourself, your business, the marketing process, and investing time with me. I want to be an encourager of you and your business pursuits.

When you see the humanness in your advertisers, and learn about them, you'll also be learning about yourself.

Relax

Mentally replay the preceding self-talk before you go into an advertiser's business. It will calm you and remind you of your purpose for this call before you get caught up in *your* nervousness and begin to focus on yourself. When you *focus* on your fear of meeting new people or dealing with difficult people, you *increase* your fear. When you concentrate on how the other person may be feeling, and focus on making *him* feel more comfortable, you let go of *your* fear. You may want to tape this mental conversation and play it back in your car just before you go in. You'll be amazed at how much more confident and relaxed you'll feel. You'll be accomplishing three objectives: You'll do what many successful athletes do before an event—concentrate on a positive outcome. You'll remind yourself of the humanness of the person you're about to approach. And, you'll relax. When you relax, you let go of the fear of failure and you're able to enjoy the process. You're sure of your reason for being there. You remember that you're good at what you do.

Many clients are grateful for the help you can give them and their businesses. When you approach people while carrying such positive thoughts, they will pick up on them— intuitively. They will be more open to you. They will feel calmer in your presence; without being consciously aware of it, people are attracted to the calmness of others. Consider all of the events and politics that could be occurring in any advertiser's life. They could use some time at your oasis of serenity. There's a bonus to this—when you are quiet within, you bring out the best in you. When you focus on helping others, you always help yourself.

The Radio Voice

One of the most obvious signs of an on-air rookie is a "radio voice." No doubt you've heard an announcer using his new on-air voice in a commercial to persuade you to buy something. When he's reading the ad in the production room, you can just imagine him thinking how *sincere* and *convincing* it sounds. The voice you hear is usually two octaves lower than a normal speaking voice. Actually, if it was able to go any lower, the rookie might be able to quit the radio job and get a better paying one as a foghorn in an extremely murky coastal area. When on-air, the radio rookie believes that it's his duty to be "cool." You guessed it. All "really cool" people have incredibly low voices. After all, I'm in *radio*, he reasons. That's what management wants from me, he believes. What the enthusiastic neophyte doesn't realize is that management *won't* be listening for the first week—or so. Due to past experience, management is guardedly optimistic that the radio-voiced beginner will have calmed down by then and will either have improved or will have laryngitis and the problem will have solved itself. Such are the fantasies of management.

The Sales Voice

Many first-year salespeople also have a "radio voice." If you've ever seen an airing of *Star Search*, you've probably seen the spokesmodel contestants. They get an opportunity to show how skilled they are when they read a promo for the upcoming segment that you'll see—right after the commercials. Most of the candidates can read fairly well, but they don't usually listen to what they're saying. There is very little feeling behind their words. Someone—their parents, a friend, or the janitor—could have suggested that they *think* about what they're saying. They're nervous, of course, and they're not professionals; but their maturity level would have appeared higher if they'd thought about what they said. One might have actually believed that the spokesmodels were really interested in what they were saying.

Similarly, rookie salespeople have been known to use their new "sales-pitch voice" when they go on calls. As the months pass, they tend to relax during sales presentations and, eventually, they lose their new "radio voice." If you're an inexperienced salesperson, invest in yourself by taping several of your presentations. Listen to yourself as if you are the prospective client. Do you sound like you really *mean* the words you're saying? Do *you* believe that the person you are hearing is sincere? Would *you* buy from you? Ask your sales manager to critique your presentations. Ask your program director to listen to your tape and make suggestions. This may be a difficult exercise, but it is worth the effort. Soon you'll sound more sincere and have more confidence in your presentation skills. Your clients will also have more confidence in you. Selling will be more fun.

How to Deal with Advertisers' Fears

Sales managers and trainers once thought it was sufficient to help salespeople recognize *their* fears in order to deal with them. Today, however, you need to go further and examine your advertisers' fears also. Many sales reps are so focused on their fear of rejection by an advertiser that they forget to look at the sales call from the client's point of view. Some salespeople report that they feel business decision makers are afraid of making *any* advertising decision for fear it will be the "wrong" one. Sometimes advertisers' fears are based on spending money on something they don't understand. Oh, they won't tell you that they don't understand the process, or the plan, or the offer, because that will bring up another fear—the fear of trusting a salesperson. Unfortunately, salespeople often add to this mistrust when they use a lot of radio insiders' vocabulary. For advertisers who are not skilled at ordering media, unfamiliar vocabulary only clouds the issue. Instead of inspiring confidence that the salesperson is knowledgeable and will be a big help in planning and implementing a successful advertising campaign, just the opposite happens.

The thoughts and feelings advertisers experience are similar to the homeowner at the greenhouse who is trying to solve his landscaping problems. He leaves the greenhouse with his head filled with so much technical information about chemical formulas for just *the right* fertilizer that the homeowner is frozen with indecision. He's afraid to make *any* decision for fear it will be the wrong one. Maybe he'll buy the wrong fertilizer and kill all of those expensive shrubs in his yard. Maybe his wife will kill him. Maybe the neighbors will laugh at him. "Better wait and think about it," he tells himself. What he really means, of course, is that maybe he should think about something else—*anything* else that doesn't give him a headache.

More Fears

The fear of *success* from an advertising buy is usually found in *managers* instead of owners. The manager may fear that if the advertising is successful and increases sales, he'll be asked to deliver this increased sales volume month after month. This thought may propel him into a zone of extreme discomfort. *Owners*, on the other hand, tend to possess the fear of *failure* from an advertising buy; if it doesn't work, how will the bill be paid or how will other obligations be met? The fear that an advertising plan will fail can lead to a much larger fear about their business failing, which could lead to a loss of independence and even a loss of identity.

An owner looks at his business and says to himself, "I own this. It's a wonderful feeling." There is a sense that no one can tell the owner what to do even if he's very aware that the bank and his customers often dictate his actions. Business owners and managers sometimes also fear the loss of employees or that their employees may unite against them. Add to this list the fear that customers won't come into the store. Then, there's the fear of competitors taking away customers to the point that the business won't survive. Owners and managers also have a personal life—most of the time. You haven't even examined the fears these people are dealing with on a personal level. As we all know, personal lives spill over into one's business life. Many, or all, of these fears may be swirling around an advertiser when you walk into a business. It's no wonder advertisers are often apprehensive about investing their time with salespeople.

Take a Number

Take the number of media salespeople an advertiser may have calling on him and add an incredible list of vendors who are trying to sell their lines to this person. The advertiser sees all of you as "salespeople." You thought you were already at the end of a long line of media people and now you need to include the vending reps. But wait, the line gets longer! All business decision makers also have amateur salespeople calling on them. Amateur salespeople are the ones who represent fund-raising events for charities or schools. You know, the students who are selling ads in the high school yearbook or the off-duty firefighters who want to sell an ad in their program for annual field days. "Oh, but that isn't real advertising," you say. It may not be "real advertising," but it is emotional P.R.; and it adds a lot of people to the ever-increasing line of "sales reps" calling on your advertiser. Now that you can see the line of sales reps queuing outside the door and around the block, what can you possibly do to help an advertiser overcome any, or all, of these fears?

Serve Others

Just as a sales manager's job is to serve salespeople, a sales rep's job is to serve advertisers. That means caring about clients as people *first* and finding out why they're in the business they've chosen. What do they *love* about their business? True marketing consultants serve their clients by helping them focus on bringing out the best in the people *they* serve (their customers). As a salesperson, look for the best in people. Expect people to tell you the truth. Open yourself to the goodness in others. When you genuinely like the people you call on, they will know it—instinctively. "Oh sure, you haven't seen some of the people I call on," you say. Well, think of one of the business decision makers you try to sell advertising to that has not been particularly considerate or even polite to you. Picture this person in your mind's eye. Now, look past her actions. What are some of the possible reasons for treating you the way she has? What could she be dealing with that you don't see on the surface?

Power. To make a terrific presentation, a sales rep was trying to gather information from a prospective advertiser. To gain the necessary data, however, the rep needed to interview the company's bookkeeper. The interview was friendly and informative; and the bookkeeper, Ms. Wilson, agreed to supply several marketing reports that were generated by her computer program. Unfortunately, the time allotted for their interview ended before the reports could be printed. The rep agreed to call Ms. Wilson the following day to arrange to pick up the marketing reports.

The next day when the salesperson called the bookkeeper, she was in a meeting. Indeed, she was in meetings the following *three* times he called that day. After many calls over the next several days, the sales rep finally connected with the bookkeeper. Ms. Wilson explained that many changes were occurring within the company and that he would have to wait until the following week to get the reports. After demonstrating what he considered incredible patience over the last five days, the sales rep explained to the bookkeeper that he understood her situation: "I know you're really busy and I don't want to add to your workload. I'll just call Jerry (the prospective client) and see if he can retrieve the reports." Hastily, and with frost on her breath, Ms. Wilson agreed that the rep should call her boss. She then hung up. The sales rep finally reached Jerry, by phone, the next day. Jerry said he would speak to Ms. Wilson and see when the reports would be ready.

Encountering the bookkeeper in the hallway that afternoon, Jerry asked about the marketing reports for the advertising rep. Ms. Wilson elaborated on the endless rounds of organizational meetings taking place. She stated what she felt were the most important priorities for the company, at the moment. Jerry remembered all the years of dedicated service Ms. Wilson had given to the company. She was the best bookkeeper they had ever employed. He quickly told her to prepare the reports when it was convenient for her. Ms. Wilson sighed and then confided, "I probably could have had the reports ready this afternoon but when the sales rep threatened to go over my head and speak with you, I simply let him." As Jerry walked away, he wondered if the sales rep really understood the meaning of power. Everyone has power, he thought. Everyone.

Perspective. From the sales rep's point of view, he had been *especially* diligent in following through with his phone calls to the bookkeeper. She, on the other hand, had moved with the speed of a dinosaur caught in a tar pit. The sales rep knew that he needed to make a presentation to his prospective client within a few days of his initial interview. As time passed, the advertiser would forget more and more of the meeting and the rep's exceptional credibility would dwindle. Unfortunately, advertisers' businesses *often* don't run like you'd wish. There is no one solution to the sales rep's dilemma. Perhaps he may have tried contacting Ms. Wilson as soon as she arrived one morning, making sure his call reached her before her meetings commenced. He may have needed to ask the receptionist when Ms Wilson arrived and what time her meetings began. In fact, he might have had to make inquiries of several people to learn her daily schedule. Actually, all of his efforts may not have rewarded him with the necessary reports on *his* time schedule. Most of life doesn't progress as planned. Seeing the world through the bookkeeper's eyes, however, might have helped.

Business Cards Can Help. Consider putting your home phone number on your business cards. Electricity may be pulsing through your body with the mere mention of such an invasion of your privacy. You may be thinking of all the hours you devote to your job now and envision that number rocketing upward if your clients were encouraged to call you at home. If you happen to be a single female you may not feel comfortable placing your home phone number on a silver platter and handing it to your advertisers. Real estate salespeople realized a long time ago, however, that they needed to be available to their clients outside of regular business hours. If you feel absolutely uncomfortable using your home phone number, print your pager or voice mail number on your card. Also print that you offer "client service 7 days a week."

Most of your clients are probably in the retail business. What are "regular" hours in retail today? Experience has shown that phone calls to your home don't increase beyond a single-digit percentage. The extra service you imply for your clients, however, helps you stand out from competitors with double-digit appreciation from advertisers. When it's time to renew a very lucrative contract with an advertiser or you're trying to close a sale, the extra effort in service may be worth a few phone calls. Wouldn't you choose to fix a potential problem over the phone on the weekend rather than walk into the station on Monday morning and find out that an unhappy client had cancelled a contract that took months to close?

Eliminating Negatives

Most people agree that there are no perfect people. Likewise, most people believe that there are no perfect jobs. Even if you love your job, there are probably some aspects of it you'd like to see vaporized and never materialize again. One way to turn some of the negative aspects into positives is to use the exercises in Exhibits 9.1 and 9.2. If money, or training, was no object, what kind of job would cause you to eagerly anticipate the beginning of each new day? How could you combine those interests with your sales position, right now?

If necessary, talk with your sales manager. Two brains looking for a solution are often better than one. Before you talk with your manager, however, give some thought to how you believe *you* can change or adapt the items on List 2 (use Exhibit 9.3).

Exhibit 9.1

List 1: What I Like About My Job

1)

2)

3)

4)

5)

6)

Exhibit 9.2

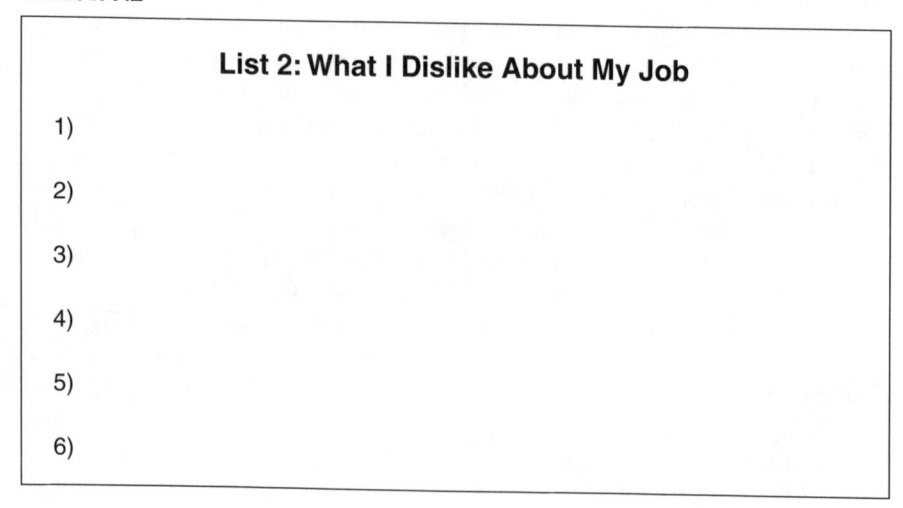

List 2: What I Dislike About My Job

1)

2)

3)

4)

5)

6)

Exhibit 9.3

What can I change on List 2? How can I change it?

1)

2)

3)

4)

5)

6)

Prioritize, Analyze, Fantasize, Then Negotiate

First, *prioritize*. Which things on List 2 do you dislike the most? Would small changes satisfy you or are these major annoyances in your life that you *really* need to adjust? Second, *analyze*. Look at the first item on your list. For example, if you detest filling out call reports, write why you dislike this part of your job. Be honest with yourself. This part of the exercise can be for your eyes only. Third, *fantasize*. How could some of the items be changed by just *your* actions? Which points involve other people—coworkers, clients, or family? Itemize your suggestions. Double check to see how realistic your plan will appear to those involved. Fourth, *negotiate*. If you feel you have a lack of autonomy, for instance, suggest ways you could interact differently with others that would allow you to meet your obligations while feeling more involved in the decision making that concerns you. For example, if you want to be responsible for deciding the number of calls you need to make daily (or weekly), explain this to your manager—but make sure your communication is nonjudgmental. Always take responsibility for your own actions, or inactions. If you're suggesting that someone else take some of your less important tasks, think through the process carefully. Would you be putting an unrealistic load on a co-worker? An unpleasant task for you might be an opportunity for someone else to shine. There may be someone who is eager to develop a new skill or show supervisors how she can handle additional responsibilities.

After you've analyzed your list and created a plan to change the things you can change by yourself, it's time to talk with your sales manager about the other items. If you're prepared, and reasonable, your manager should consider your ideas and even add suggestions. Your sincere effort will increase your chances of changing some negatives in your life to positives.

Negotiating for "The List"

When a salesperson begins a new job at a radio station she usually inherits an account list. The size and the perceived quality of the list vary. The negotiation for "the list" is seen as an important aspect of getting that new job. Are there a lot of advertisers on the list who are already on the air? What percentage of the advertisers are on-air *every* month? Regardless of the outcome of the negotiations, approximately 40% of that list will change during the next 12 months. Why should an account list that was so important during the job interview process change so much within such a short time? Well, certainly, some businesses don't survive. A number of businesses are sold to a new owner who is totally convinced that he has a fail-proof method of advertising, which doesn't include radio. These two categories together, however, usually don't account for 40% of a list within a year.

So why would such a large portion of an account list change so quickly? The reason is the sales rep. To be more exact, the sales rep's personality and interests. People tend to excel in a particular area when they are interested in it. It isn't that the sales rep doesn't attempt to sell to everyone on the list, it's just that she'll be more successful working with businesses that sell products and services she would personally enjoy buying and using. The buyer's personality is also a factor. Anyone who has been on the street selling for very long knows that you need to highlight traits that you possess to match the personality traits of *each* buyer. Sometimes, however, the match isn't very strong or just isn't there. Then, to maintain billing, you must begin a search for advertisers you can relate to more successfully. At this point, salespeople begin to recreate their account list. Usually this process is painful and takes an enormous amount of time. Here are some guidelines that may make the procedure easier for you.

Creating a Dynamic List

Look for businesses that sell products *you'd* like to own. You'll be the most effective when you direct your energies toward the kinds of businesses you enjoy buying from—and you'll have more fun. What kind of accounts do you want to call on?

- What types of products and services do you enjoy buying and using?
- Do you possess a talent or skill that may or may not be used on your job? Do you use that talent or skill to pursue a hobby?
- Do you require special equipment for your hobby? In other words, would you enjoy working with businesses that may be connected to your hobby?
- Are you a homeowner?

- Do you invest in home electronics? Are you an audiophile who loves stereo equipment?
- Do you live in a condo? How did you find your condo?
- Which needs were you trying to fill when you were searching for a place to live?
- What personal goals are you working toward?
- Do you want a new car?
- Are you saving money to buy new carpeting?

You are a consumer; there are many other consumers who share some of your needs and interests. You already have a head start when you pursue businesses that interest you because you may have gone through the buying experience or have given it a lot of thought. You know some of the concerns and questions their customers ask.

When you're choosing businesses on which to call, select ones you believe will do well on your radio station. Think about the process each consumer group goes through before buying from your prospective clients. You don't have to know all the specifics of buying patterns before you approach a prospect because niche marketing analyses will give you many of those answers. Just answer the questions here as honestly as you can, build a target list, and begin conducting niche marketing analyses.

Choosing Your Fears

You choose your fears. You decide what, or who, you will fear. Then, you decide which fears you'll face and defeat and which fears you'll continue to keep in your life. Of course, what is fearful to one person may be a joy for another. You decide what your fears will be. Then, it's your job to face them, with or without help. Almost every day someone is asked to do something that provokes fear. Whether it's calling that mean and nasty advertiser who gives you a hard time *every* time you go in to see him or it's writing an ad for a client who is impossible to please.

When you're confronting your fears, allow yourself the luxury of feeling the fear. It's OK to admit to yourself that you're afraid. Examine your fears one at a time—look at a particular fear from all angles. Ask yourself to mentally list all the possible outcomes if you were to confront this fear. If you can think of only negative outcomes, keep searching. When you have mentally drawn sketches for each scenario, tabulate how many outcomes are not positive. What are the imagined obstacles? Remember, these are perceived obstacles. Picture each one in detail; then, see each one exploding and disappearing. The remaining positive outcomes—usually in the minority—can be studied and turned into opportunities. This process doesn't *guarantee* a positive outcome; but if you focus on a positive result, you're much more likely to achieve it. After you choose your fears, choose to face them. Only then will you feel better about yourself. Approach that businessperson you've been avoiding. Search for new ways to contact the prospect who's been so elusive. Take a chance on yourself.

Risk Today

Every time you avoid taking a risk, you're saying to yourself that you can't handle the outcome. You're capable of much more than you believe you are, however. Take at least small risks every day. You'll be expanding the previous limits of your comfort zone. Taking risks will become an accepted way of life and you'll be pushing past prior *self-imposed* limits. All limits, after all, are self-imposed. When you become accustomed to taking risks in one area of your life, you will begin to take risks in other areas as well. You'll develop an ability to break through previous fear-based limits and take risks that help you realize the fulfillment of your dreams. You'll feel wonderful about yourself!

Comparisons

Stop comparing yourself to other salespeople. You haven't had an identical life, how could you possibly have an identical selling career? If you must compare, compare yourself to where you were one month ago, six months ago, one year ago, or five years ago. In the history of the universe, there has never been another person just like you and never will be again. No other human being has had your exact experiences in life. You possess unique gifts that only you can give to the world. It is ludicrous to compare one life to another. It's much more productive to compare your triumphs today with your past actions and achievements. Look at what you've accomplished!

Breaking Through

When you become unhappy in a situation, when you become depressed or feel you can't bear the situation any longer but don't know how to remedy it, you are about to break through to a higher plane of understanding. For example, a salesperson at a broadcast station was unhappy in his job. The sales rep was searching for a sense of team spirit; but regardless of his efforts, the rep felt separated from the sales staff. He had considerable success in sales, but was not feeling very confident with his new clients. He also felt a lack of support from his sales manager. The salesperson continually questioned and lamented over his apparent inabilities. His acceptance of a need for change came when one of his clients wanted a lower rate than they had been receiving. The salesperson negotiated but refused to lower the rate any further. He discussed his client's demands with his sales manager. Without informing her salesperson, the sales manager phoned the client and agreed to lower the rate.

When the salesperson learned what had happened, he realized the atmosphere at the station, and the lack of good relationships there, were not conducive to living a happy and fulfilled life. He recognized that the situation wasn't really about his ability to sell or to bond with other salespeople. The rep realized he had a limit to what he would do for his career or for money. He weighed the value of career advancement and money against peace of mind and belief in himself. In that moment, his vision became crystal clear. He broke through to a higher plane of self-awareness. The next day he went into the sales manager's office and resigned. He was apprehensive about the future but confident in his ability to survive. He was at peace with himself for the first time in a long time. He exhaled.

Letting Go

It's lonely out there on the street. You're walking the sales gangplank every time you ask someone to buy from you. It's only when you let go and disregard the chance to fail that you accomplish your best work. Your best sales often happen when you give up. When you say to yourself, "I can live without this sale. I have nothing to lose. I *must* tell this person what is really on my mind. I need to look back on this moment and say I gave the best business and marketing advice I could. If this advertiser *never* does business with me, I can feel good that I said the truth, as I see it." You should feel that you can look back at this moment and say, "Yes! I gave it my best." So let go and trust yourself.

Affirmations Can Help Change Old Habits

The pictures on your walls in your home and office have a subtle influence on your thoughts. Usually you've chosen those pictures because they give you pleasure. Words can have an equally subtle, and sometimes more profound, effect on you. Affirmations are positive and optimistic declarations usually written in the present tense. They serve to remind you of constructive and beneficial thoughts, which can help you during the everyday challenges of your life. The list of affirmations in Exhibit 9.4 may appeal to you or you may choose others. Place affirmation lists in your bathroom, your kitchen, your office, and even your automobile where you'll see them every day. Print or type the affirmations large enough so you can read them from across the room. If you feel you need more specific affirmations, look for phrases that have special meaning for you. When you begin to look for them, you'll find affirmations that are appropriate for you. If you place some affirmations on or near your desk, you'll find that your eyes will go to the list just when you need it most. Sometimes, your eyes will be drawn to just one line, but you'll find *that line* pertinent and very comforting at that moment. Try it.

Exhibit 9.4

> Be patient with yourself, (*your name*).
> Slow down.
> Let go. . . . Trust yourself.
> Remember what it feels like to feel good. . . . Enjoy.
> Don't be afraid of *New*. . . . It might be fun.
> Do what feels good.
> Just take care of today.
> You are more than you believe you are.
> Enjoy the moment.

Psychologists teach that to replace an old habit you must repeat a new action a minimum of 21 consecutive times. Because your mind moves toward and works on your most prominent thoughts, what you think about *most* tends to happen. When your most dominant thoughts focus on an advertiser saying *no* or your sales manager being upset with you because you haven't reached your quota, you tend to realize these occurrences. Because most people's thoughts (approximately 85%) are inclined to be negative, you'll need to input positives on a daily basis. Some experts caution that if you don't program your subconscious on a daily basis, someone else will. You're in the media business. You can see firsthand the effects of subconscious programming on consumers.

When you write affirmations for yourself, use the active tense of a verb. For example, instead of saying "I am going to lose weight," use "I am losing weight *now*." Picture your new slimmer self. How do you feel? How do others respond to you? How do your clothes feel? Do you have more energy? Do you think more positive thoughts about yourself? Each time you look at your "losing weight now" affirmation, answer all these questions in your mind. You'll be affirming that you will achieve your goal.

As you reprogram your subconscious, it will become easier to perform your new, more productive habits. Apply this technique to finding more advertisers who want to buy advertising from you. See yourself reaching new heights in your career. Imagine being even more successful than you have previously fantasized. You'll be programming your subconscious to propel you beyond previous limits you may have unconsciously created. Ready. Set. Imagine.

Giving Encouragement

How many years of your life do you work? What percentage of your life does that represent? It's less than the blink of an eye from the perspective of the age of the universe. Most immortality is produced from an idea, an encouragement you give to someone who is unsure of himself. You seldom know how your words and actions may have impacted someone's life. Like the ever-increasing rings on water after a stone is dropped into it, encouragement goes out in little ways touching many people. Think of the times someone has given you encouragement—just when you felt you needed it the most. One of the ways you can repay the people who have helped you is to help others. Like small waves continually rolling onto the shore, give encouragement to everyone—your family, your co-workers, your clients, strangers, even your supervisors. Pass it on.

Advertisers as Friends and Family

Some of the most successful salespeople don't call on advertisers. These salespeople go into businesses to see friends and talk about their friend's favorite subject—his business and how to make it more successful. No doubt you've heard someone say that there are no strangers, just friends you haven't met yet. You've probably also heard that you need to treat your friends like family and your family like friends. These two pieces of advice may seem overused or even corny, but they still remain words of wisdom. Think about it for a minute.

Think of a particular friend of yours. Does visiting with her give you pleasure? Are you anxious to hear about events happening in her life? Are you interested in her perspectives—even if they're different from yours? Do you share *your* experiences with her? Are you eager to learn as much about your friend as possible? Do you support her when she makes decisions about her life—even if you may not personally agree with those decisions? Do you accept your friend for who she is without trying to change her? Do you think about your friend even when you're not together? Do you analyze her challenges, looking for ways to help her? Do you find being with your friend effortless? You probably answered yes to these questions. Can you go back and replace the vision of your friend with the picture of an advertiser with whom you work? If you can truly like the people on whom you call, you'll embrace your job with a self-renewing vigor. You'll want all the positive things for your advertiser that you wish for your friend. Take a risk and make the effort to turn your advertisers into genuine friends. You may find yourself in love with your job. Success will embrace you.

Preparing to Create Wealth

Someone once said that to gain wealth you simply have to do what wealthy people do. That sounds simple enough. But what exactly do wealthy people "do"? Many people who have gained wealth first became very clear in their mind about what they wanted in exchange for their time and efforts. Then, they examined their conscious and subconscious feelings about having money. They also began by believing that they deserved wealth. Many people who believe they want large amounts of money often sabotage their efforts. On a conscious level, they have very different views about accumulating and owning money than they do on a subconscious level. Everyone has programmed their subconscious with beliefs about themselves that began in childhood. Your subconscious doesn't make a judgment about the ideas you store in it. It just files *all* information as if it is reality. Until you're able to examine beliefs from the perspective of an adult, you may not know your true feelings about accumulating wealth. When you're unclear about what you are willing to give up in order to create wealth for yourself, conflicts arise. When there is a conflict between your subconscious and your conscious mind, your subconscious always wins.

Take a sheet of paper and draw a line down the middle, from the top of the paper to the bottom. At the top of the paper on the left side, write the words "I deserve wealth because . . ." On the right side of the paper, write the words "I don't deserve wealth because . . ." Take a deep breath and exhale very slowly. Don't push for the answers but begin to write. When you've finished this exercise, get another piece of paper. At the top of the paper on the left side, write "When I have wealth, these positive things will happen to me (and my family)." On the right side of the paper, write "When I have wealth, these negative things will happen to me (and my family)." Again, a relaxing breath will help you begin to bring information from your conscious mind as well as your subconscious mind. When the negative thoughts start to surface, don't deny them or try to analyze them at this point; simply record your thoughts. You may be very surprised by the information that emerges.

When you've completed your writing exercises, carefully read your answers. When you study the negative responses, don't try to deny the words you wrote. Instead, look at each response from the perspective of an adult. Was this possibly information that someone else put in your subconscious based on *their* beliefs? As you examine this data, decide if you agree with it or if you need to study it and reprogram your subconscious to correspond with beliefs you've decided to embrace as an adult.

A Sitcom

Life is not a 30-minute sitcom where all problems are solved in the 29th minute. Allow yourself *time* to become the person you want to become. The only reality you truly know is *now*. Yesterday is gone. Tomorrow doesn't exist. Focus on this day and simply do what feels right for you today. Handle one day at a time, focus on just one call at a time. Be patient with yourself. Someone once explained that to become an outstanding tennis player one only has to serve well—consistently. To become an outstanding niche marketing consultant, you need to verbally serve well—consistently. Knowledge is the key. A thorough knowledge of your medium is uppermost on your list, of course, but a superior knowledge of other media is also necessary today.

Understanding your advertisers' fears and seeing *their* world through *their* eyes is an important segment of the pathway to knowledge. The most important ingredient in this mixture, however, may be self-knowledge. Learn from every resource available to you—your prospects, your clients, your co-workers (*all* your co-workers), your supervisors, books, magazines, tapes, seminars, your children, and *anyone* who'll share life experiences with you. In other words, there is no time-out. *Everyone* you encounter can give you the gift of their perspective. Relish their gifts.

PART FOUR
Presentations and Niche Copywriting

10

A Niche Presentation for an Auto Dealer

Great Books Versus Great Presentations

Broadcasters love to receive great books such as the *BBM* in Canada and the *Arbitron* in the United States. They wait for magic numbers to bring them good fortune and a little fame. Today, fragmenting markets and the explosion of media choices are forcing the radio industry to seek a different kind of great book and a radically different approach to media sales.

Presentations by Fax

In an age of modems, computers, and fax machines people tend to be brief, concise, and to-the-point. "After all," argue many radio managers, "if our civilization is suffering from an information overload, we need brevity!" However, when you focus on small snippets of information to help you sell your product (your station's air time), it becomes increasingly difficult to establish value. Presentations often dissolve into rate negotiations, or "value-added" concessions, which usually translate into bonus ads.

Stop the Insanity

You can choose to stop the insanity. You can decide to make logical, consumer-focused marketing presentations to your advertisers and potential advertisers. The following presentation was made to a media planner at a regional ad agency. The names have been changed to protect the marketing strategies of the specific business. The station for which the rep worked was mainstream adult contemporary. There were three other AC stations in a market with 26 radio stations. It's not easy to stand out in a radio crowd that large.

The Assignment

An ad agency media planner needed some promotional ideas for May through September for an auto dealer client. The radio station's sales manager asked Hope, the sales rep, to see if the client would consider giving away an Isuzu Trooper. The media planner phoned the radio station at the last minute and told Hope to meet him at the dealership. The rep presented to both of them in this meeting. Because she had expected to present to only the media planner, Gene, she didn't have two copies of each sheet in her presentation. Like most radio salespeople, she was flexible, however, and presented the media planner first and then addressed the client. She wanted to acknowledge the client without appearing to ignore the media planner. Did she make the sale? Listen to the presentation before you learn the outcome. Words in quotes will represent the verbal portion of the presentation made by the salesperson.

Getting Started

"Please take a moment to look at my introduction."

The media planner is handed the cover sheet (Exhibit 10.1). Gene looks at the paper and passes it to the client.

Exhibit 10.1

Presentation Cover Sheet

Making More Money for STAKES ISUZU in 199__

A Special Presentation

developed for

GENE KIRKMAN

Media Planner

ADS UNLIMITED

Prepared by:
C. Hope Niles
Marketing Consultant
20 April 199__

Niche Marketing Intro

"Thanks for investing this time with me this afternoon. As you well know, there are a lot of radio stations in this market. With all the choices available, how do you make the best decision for Stakes Isuzu? We know that reaching THE MASSES isn't very efficient or cost-effective. Using niche marketing techniques will help you narrow the field. Let me show you our niche."

The rep gives the next sheet (Exhibit 10.2) to the media planner. He reads it and passes it to the auto dealer.

Exhibit 10.2

Advertising Decisions Involve Intelligent Choices

With all the advertising choices available in the Memphis SMSA, how do you make the right decision for STAKES ISUZU?

Today "Mass Communications" is an oxymoron. With the fragmentation of lifestyles . . . reaching "the masses" is not good enough.

Welcome to the New Era of Niche Marketing

or

Finding Riches in Niches

Welcome to the Station

"I'd like to welcome you to a unique radio station. At ABCD we have very carefully chosen an audience with great buying power. During the next few minutes, I'm going to show you how easily you can reach this affluent, 25–44-year-old, adult market of consumers who have the greatest potential for buying Isuzus."

The rep hands the radio station welcome sheet (Exhibit 10.3) to the media planner. He gives it to the client after he's read it.

Exhibit 10.3

WELCOME...

To the RADIO STATION that delivers the audience with the best buying power for STAKES ISUZU.

To Niche Marketing with ABCD. Invest a few minutes and discover how easily you can reach the AFFLUENT 25–44 adult market in the Memphis SMSA . . . for a very modest investment.

The Prime Market Page

The rep turns the next page upside down, places it in front of the client and media planner, and shows them the demographic breakdown of the market.

> "As you can see, according to Duncan's *Radio Market Guide*, the Memphis market breaks down like this. People aged 25 to 44 represent 40.2 percent of the market."

The rep circles the 25–44 demo with her marker and then pushes the paper to the planner (Exhibit 10.4).

Exhibit 10.4

Memphis, Tennessee
THE PRIME MARKET
ADULTS 25 to 44

According to Duncan's *Radio Market Guide 199__*, the Memphis market breaks down as follows:

Persons 12 to 34:	27.0%
Persons to 25 to 44:	40.2%
Persons 45+:	32.8%

According to the Topscan Rating Analysis System, the Memphis metro survey area's estimated population for adults 25 to 44 is 232,100.

Profiling Your Listener

"Our 25- to 44-year-old audience represents over 40% of the overall buying power in this market! This is an acquisitive group of adults with above-average needs for automobiles—Isuzus, household goods, clothes, food, and a lot more. These people need a lot of stuff! And, they have the purchasing power to buy it! Over 40 percent of our listeners earn over $35,000 annually! They can definitely afford your Isuzu! Our information is taken from *Media Probe* for the metropolitan statistical area of this market."

The media planner is handed the prime market page (Exhibit 10.5) that follows. After Gene and the client read the sheet, Hope continues.

Exhibit 10.5

ABCD Radio's PRIME MARKET

ABCD's primary target . . . adults 25 to 44 . . . represents over 40% of the overall market buying power!

This is an affluent, acquisitive group of adults with above-average needs for:

AUTOMOBILES

Household goods and services

Clothing

Food . . .

Over 42% of ABCD listeners make in excess of $35,000 yearly!

Over 78% are college-educated!

Source: Media Probe, Memphis SMSA, Qualitative Report, 199__.

Who Buys Imports?

"I'd like to take a few minutes now to focus on people who buy import vehicles and how to get them to buy *your* Isuzu. From a 199_ *Newsweek* magazine survey, it was reported that the *average* buyer spends almost *three-and-a-half* weeks before buying a vehicle. And, they visit over *four* dealers, on average. Do you find those figures are consistent with what your salespeople hear?"

The sales rep waits for feedback and then continues:

> "The *Newsweek* survey also found that only 19 percent bought from the first dealer they saw. Do you have any idea what percent of first-visit buyers your salespeople close?"

They all discuss the closing ratios of the dealer's salespeople and first-time buyers.

Loyal Buyers

> "According to a *Wall Street Journal* survey, there's good news about brand loyalty. They found that once sold, *47 percent* will buy the same make again. Men seem to be more loyal than women because *43 percent* of women and *51 percent* of men stick with the same brand. That's a great reason to court these buyers and to ask them to come back to STAKES again and again. Do you find brand loyalty high for Isuzu buyers?"

The client and the media planner confirm a high loyalty among Isuzu buyers.

Import Buyers

> "Let's take a look at the ages of import buyers. ABCD's core audience represents *43 percent* of the age group of import buyers! The 25 to 44 age range represents almost *half* of all buyers! Is STAKES ISUZU getting its share of the 25- to 44-year-old buyer?"

After the client and media planner respond, the media rep circles ABCD's demographics and hands Exhibit 10.6 to the planner and says,

> "This is what we've already discussed. You may choose to keep it for your records."

Why Consumers Buy

> "Let's take a look at why they buy in the first place. R.L. Polk surveyed import buyers and found that *durability* was the most important factor in buying. They found that many buyers feel that imports tend to be more reliable and less expensive than the American-made counterparts."

The sales rep places the page in Exhibit 10.7 in front of the media planner so he can see the list of factors in buying.

> "As you can see, the number two reason for buying is value, and then price. If these are the three top reasons for buying imports, and R.L. Polk says they are, then we suggest that your advertising focus should include themes that correspond with the top three reasons for buying. Does this make sense to you?"

Exhibit 10.6

SHOPPING AROUND

The average buyer spent 3.2 weeks visiting 4.2 dealers . . . before buying.

19% bought from the first dealer they saw.*

LOYAL BUYERS

According to a *Wall Street Journal* brand-loyalty survey . . .

47% usually buy same make again . . .

51% men

43% women

Demographics

Sex: Male . . . 56%   Female . . . 44%

Age: 18 to 24 — 8%
     25 to 34 — 19%
     35 to 44 — 24%
     45 to 54 — 16%
     55 to 64 — 15%
     65+   — 18%

Source: 199_ Newsweek survey.

The rep waits for a reply and then directs her attention to the next page of the presentation.

After They Leave the Lot

"Now that we've looked at *why* consumers buy imports, let's see *how* buyers use them after they leave the lot. According to the 199_ *Newsweek* survey we discussed earlier, the number one answer is that *61 percent* of buyers say that they use the cars for commuting. That's certainly a significant amount! *Twenty-two percent* report that their vehicles are used for local travel. *Seven percent* say they use the cars for business only. And, *6 percent* use their imports for pleasure only. Do these figures match yours?"

The client and media planner agree that their figures are very similar.

Exhibit 10.7

199_ Import Car–Buyer Survey
Most Important Factors When Buying

1. Durability
2. Value
3. Price
4. Dealer location
5. Financing

Many car buyers feel that imports tend to be more reliable and less expensive than the U.S.-made counterparts.

Source: R.L. Polk, *Consumer Buying Factors*, 199__.

Marketing Guidelines

The sales rep goes on to encourage a conversation comparing the survey results with the dealership's experiences.

> "This study also found that an overwhelming number of buyers—*67 percent*—financed their purchases and only *33 percent* paid cash. When we know the predominant reasons for buying your vehicles, then, we understand how people use those vehicles, and we know how buyers pay for their purchases, then we can use this information to make your advertising and marketing efforts more effective. These answers give us a *guideline* to follow when creating your ads. These are the statistics we've just discussed."

The rep gives Exhibit 10.8 to the media planner and says

> "Are there any questions or comments about anything that we've discussed so far?"

Hope responds to questions or comments before she proceeds.

RAB

> "I understand that you're interested in some marketing and promotional ideas for May through September of this year. ABCD is a member of the largest trade organization for radio stations. It's called the Radio Advertising Bureau, or simply, the RAB. This organization is very important to our radio station and to our advertisers. Many of the marketing suggestions we present to our clients originally came to us from radio stations in other markets through the RAB. The promotions we present

Exhibit 10.8

What Asian Imports Are Used for After They Leave the Lot

61% of those buying Asian cars use their vehicles for commuting

22% are used for local travel

7% are used for business

6% are used for pleasure trips

Paying for Imports

33% paid with cash

67% were financed

Source: 199__ Newsweek survey.

have been used with radio station clients in other parts of the country. They've already been tested. The bugs have been worked out of them. Often, we have access to the phone numbers of the participants. We can call them to discuss a specific promotion. The May promotion that I'm going to tell you about next came from the RAB."

Your Greatest Growth Area

"According to J.D. Power & Associates, women represent over 24 *percent* of principal light-duty-truck drivers. Women also represent the greatest growth area for light-duty trucks. ABCD is proposing a series of STAKES ISUZU Ladies' Nites. Over the air, we'll invite women to stop by STAKES *after work* on Thursdays during May and test drive any four-wheel-drive ISUZU they choose. Some women may feel a little uncomfortable so we'll give them a special reason to visit STAKES. The atmosphere here will, of course, be laid-back with no hard sell. Your service people may give mini-seminars on rotating tires, on changing oil, and even on preventive maintenance. We'll broadcast live from here from 4:30 to 7:30 PM. Janet Walsh, our traffic reporter, will be a great spokesperson during STAKES Ladies' Nites. We'll provide child care while your salespeople are helping with test drives and demonstrations. We'll also bring refreshments and prizes."

Building Awareness

"Your Ladies' Nite remote broadcasts will help build top-of-mind awareness with female buyers. You'll be filling a marketing niche for your dealership that represents one of your greatest growth areas. Since ABCD serves the upscale working women you need to attract, a series of Ladies'

Nites makes a lot of marketing sense for you. How do you feel about this marketing effort?"

The client tells Hope that he has thought that perhaps they have needed something to target more female customers. The dealer rep asks for more details.

Any Objections?
The advertising rep is trying to uncover any objections before continuing. There are no objections at this point. She continues to describe everything that is included in the series of remotes. She tells why a series of remote broadcasts is a better investment than just a one-time event and patiently answers all questions about the May proposal. Then, she gives a copy of

Exhibit 10.9

Suggested Promotions for STAKES ISUZU . . .
May Through September 199__
STAKES ISUZU Ladies' Nite . . .

ABCD invites women to visit STAKES ISUZU after work and test drive the four-wheel-drive Isuzu of their choice (a vehicle they might feel awkward testing ordinarily). The STAKES atmosphere is, of course, laid-back, no hard sell. STAKES has made it easy for women to ask questions about their favorite trucks. ABCD will provide child care, refreshments, and prizes.

STAKES representatives will give mini-seminars on service, such as preventive maintenance, rotating tires, changing oil, and other basics.

ABCD will broadcast live from STAKES between 4:30 PM and 7:30 PM on Thursday evenings in May. Janet Walsh, the traffic reporter, will make a perfect spokesperson for STAKES during Ladies' Nite.

Ladies' Nite will gain recognition and trust from female prospects . . . putting STAKES'S name foremost in their minds when they're ready to buy!

You'll receive: The ABCD Mobile Studio
Three (20') high-visibility ABCD banners
Child care during the remote broadcast
Refreshments and prizes
20 60-second commercials promoting Ladies' Nite at STAKES
12 live announcer liners promoting Ladies' Nite
PLUS . . . 6 live call-ins from STAKES on Ladies' Nite

Each ABCD Ladies' Nite Live Broadcast delivers 38 commercials and enormous promotional extras for only $2,800 net per remote.

Note: Janet Walsh's participation does not include her talent fee, which is payable at the remote site.

the remote broadcast page (Exhibit 10.9) to the planner. He looks at it and gives it to the client.

Great Trooper Giveaway

"Next, we've created a high-profile, 15-week, marketing promotion designed to keep STAKES ISUZU's name on people's minds. From Memorial Day to Labor Day, consumers can register to win an Isuzu Trooper at your dealership or on-the-street when our ABCD street reporter hears ABCD playing on a radio. STAKES will receive 150 live call-in street reports from the ABCD roving street reporter! ABCD will be reporting from the prize Isuzu Trooper—which will have very visible STAKES and ABCD signs on Memphis streets! You'll be mentioned in EVERY report during this INTENSIVE campaign! And, STAKES will be mentioned in a minimum of 525 promotional ads during this time! Promotional ads will tell listeners to go to STAKES ISUZU to register. We'll provide removable vinyl lettering to decorate the Trooper.

The on-air giveaway of the STAKES ISUZU TROOPER will be on the Tuesday morning after Labor Day on the Mark Gardner show. We'll be happy to arrange a special award ceremony with Mark Gardner and a representative of STAKES to make the official presentation to the winner. We'll have a photographer present. Pictures will be sent to the winner, STAKES, and ABCD. You may want to print your picture in a newspaper ad since we suggest that you focus on the promotion in all the media that you use. We realize that integrated marketing efforts are the most productive. You'll find that this promotion will work well with all media you use to promote STAKES."

Diary of a Customer

"There is one other very important aspect of this promotion that we haven't discussed. When consumers come into STAKES to register for the Trooper, they'll fill out an entry form. This is a perfect opportunity for you to get a lot of information about that perspective buyer. We'll develop a customized registration form just for STAKES. You tell us what you'd like to know about your customers and prospective customers. We'll create and print your custom entry blanks. We'll even tabulate them for you weekly! You'll receive all entry forms as soon as they're tabulated so your salespeople can use them for contacts. You may also want to update your database. Here's a sample registration form" (Exhibit 10.10).

What Do Customers Want?

"You'll notice that we ask the usual name, address, and age-range questions. Then, the next paragraph asks the entrant to list the year and model of his or her present vehicle plus any second and third vehicle—if he or she owns one. In our example, we ask which car or truck the entrant would *like* to own and the features that would be important.

Exhibit 10.10

REGISTRATION *(ABCD Radio Logo)*

PLEASE FILL OUT THIS FORM AND DEPOSIT HERE. ENTER AS OFTEN AS YOU VISIT THIS LOCATION. YOU MUST BE 18 TO ENTER.

Name: _____
Street address: _____
City: _____ State: _____ Zip: _____

Present vehicle(s): Year _____ Model _____ _____
Second: Year _____ Model _____ _____
Third: Year _____ Model _____ _____
Car (truck) I would like to own: _____ _____
Features that would be important to me: _____
_____

Price range: _____ Affordable monthly payment: _____
I ☐ have ☐ have never leased a vehicle.
I am ☐ looking ☐ not looking for a _____ car or _____ truck.
I ☐ have ☐ have not purchased a _____ from this dealership.
My age range: ☐ 18–24 ☐ 25–34 ☐ 35–44 ☐ 45–54 ☐ 55–64
☐ 65–74 ☐ 75+

We've included a price range and a monthly payment with which the new owner would feel comfortable. We ask about leasing. We also ask if the entrant is in the market for a vehicle and if he or she has purchased from your dealership before this visit."

Prequalify Prospects

"With this information, you'll be able to prequalify prospective customers *before* your salespeople approach them—over the phone or in the mail. Imagine how helpful it will be for your salespeople to know the kind of vehicle a person is interested in before they call them. The sales rep will be able to tell the prospect about inventory that would specifically interest that person."

Heavy Users and Closing More Sales

"Your sales manager will be able to determine the heavy users, or people who are most likely to purchase one, two, or even three vehicles from you. Your manager could decide in advance which salesperson would work best with specific prospects.

When you see that a buyer's monthly payment amount is not in line with the type of vehicle he prefers and you find that he has never leased an automobile, in this instance, leasing may be the answer to closing the sale. Think how much more effective your salespeople can be if they have this information *before* they contact the prospective buyer. Here are the details of the promotion."

The sales rep hands the next page (Exhibit 10.11) to the media planner.

Exhibit 10.11

The Great ABCD and STAKES Isuzu Trooper Giveaway!

From Memorial Day to Labor Day, people can register to win an ISUZU TROOPER at STAKES . . . and register "On the Street" when the ABCD street reporter hears ABCD playing on a RADIO.

STAKES will receive 150 live call-in street reports from the ABCD roving street reporters! ABCD will be reporting on traffic conditions from the "prize" ISUZU TROOPER, which will sport highly visible STAKES and ABCD signs on Memphis streets! STAKES ISUZU will be mentioned in EVERY report during this intensive 15-week campaign! **STAKES ISUZU will also be mentioned** in EVERY live promotional ad for the Street Patrol *all summer long!*

STAKES ISUZU'S Traveling Billboard

The ABCD–STAKES ISUZU TROOPER will be displayed at STAKES during the week. **Promotional ads will tell listeners** to go to STAKES to register to win this high-profile vehicle!

ABCD will provide removable vinyl lettering to decorate the TROOPER for the promotion. The on-air giveaway of the TROOPER will be on the Tuesday morning after Labor Day on the Mark Gardner Show. Winners will be presented the TROOPER's keys at a ceremony at the STAKES showrooms.

Total Promotional Value: $20,250*

*Note: Includes promotional announcements, street reports, sign value, live on-air presentation, street reporters, and cross-promotion ads!

Effective Scheduling

"We've discussed why customers buy imports and what they do with them after they buy. We've discussed ways to find out why people buy from STAKES ISUZU and what will bring them back. We've looked at guidelines for formulating a good creative. Now, let's look at the last ingredient of a good marketing campaign—effective scheduling. In 1994, a second edition of the book *Radio Advertising's Missing Ingredient: The Optimum Effective Scheduling System* by Pierre Bouvard and Steve Marx was published by the National Association of Broadcasters, or NAB. The authors showed how the radio medium is effective and efficient when bought correctly. After formulating a creative, scheduling is seen as the critical other half of getting a profitable return on your investment. How many times does a commercial have to air to get results? The optimum effective scheduling (OES) system was designed to reach the majority of a station's cumulative audience at least three times in a week. Consumers contacted three or more times are labeled *effective reach*. This is part of the explanation for OES."

The media rep hands Exhibit 10.12 to the media planner. The rep waits for the planner and the client to look at the OES introductory page before continuing.

Exhibit 10.12

Optimum Effective Scheduling

Can advertisers afford to be *without* OES?

"While good creative is a necessity, scheduling is seen as the critical *other* half of the solution to generating *cash register results*."

Effective Reach

"OES has been designed to reach the majority of a radio station's cumulative audience *three or more* times in a week."

"Consumers reached three or more times are labeled 'effective reach.'"

Source: *Radio Advertising's Missing Ingredient: The Optimum Effective Scheduling System* (2nd Ed.) by Pierre Bouvard and Steve Marx, NAB, 1994.

What Is Effective with ABCD?

"What is the optimum effective schedule for ABCD? To determine the answer to that question, we look at the number of people who make up our cumulative audience and divide it by the average number of people listening to our radio station in a given quarter-hour. This gives us the turnover ratio for our audience. Then, we simply multiply our turnover ratio by the OES spot ratio of 3.29. On ABCD that translates to 44 ads aired Monday through Sunday between 6 AM and midnight. Using our audience size, that would be 29,050 consumers effectively reached with your message a minimum of three times. Using the optimum effective scheduling system makes your advertising an investment instead of a purchase. Here's the formula for computing OES" (Exhibit 10.13).

Exhibit 10.13

Determining Optimum Effective Scheduling

1. Cume divided by AQH = Turnover Ratio

2. Turnover Ratio × OES Spot Factor (3.29)*
    13.1 × 3.29 = 44 units (Monday–Sunday,
    6 AM to Midnight)

3. Cume × OES Reach Factor (.46)* = 29,050 consumers
    Effectively Reached . . . a minimum of three times*

**O.E.S. Is an Investment—
Not a Purchase**

"The success of hundreds of advertisers makes it clear that running anything short of OES cannot be relied on to accomplish marketing goals."

*These constants were produced by the Katz Probe system.
Source: *Radio Advertising's Missing Ingredient: The Optimum Effective Scheduling System* (2nd Ed.) by Pierre Bouvard and Steve Marx, NAB, 1994.

Sample OES Schedules

"Here are samples of optimum effective schedules for June through August (Exhibit 10.14). During September, we suggest another series of Ladies' Nite Remote Broadcasts."

Summary, THE Question, and the Close

"To summarize, we suggest a series of Ladies' Nite Remote Broadcasts in May and September. These will enhance the Trooper Giveaway during that 15-week period between Memorial Day and Labor Day. Are there any questions?"

The client and media planner ask several questions and then they ask THE question: How much will this cost? Hope summaries:

Exhibit 10.14

Sample Optimum Effective Schedule
for June through August 199_

	MON	TUE	WED	THU	FRI	SAT	SUN
6 AM–Midnight	7×	7×	6×	6×	6×	6×	6×

Suggested O.E.S. Flights . . .

Week #1 — Start date:	1 June	End date:	7 June
Week #2 — Start date:	15 June	End date:	21 June
Week #3 — Start date:	29 June	End date:	5 July
Week #4 — Start date:	13 July	End date:	19 July
Week #5 — Start date:	27 July	End date:	2 Aug
Week #6 — Start date:	10 Aug	End date:	16 Aug
Week #7 — Start date:	24 Aug	End date:	30 Aug
Week #8 — Start date:	31 Aug	End date:	6 Sep

* * * * * * * * * * *

Week #9 — 8 Sep . . . Live Broadcast—Trooper Giveaway

* * * * * * * * * * *

Week #9 — Ladies' Nite Remote Broadcast schedule
Week #10 — Ladies' Nite Remote Broadcast schedule
Week #11 — Ladies' Nite Remote Broadcast schedule
Week #12 — Ladies' Nite Remote Broadcast schedule

"Let's see, the giveaway promotion includes 525 promotional announcements, 150 street reports, the value of a traveling billboard, a live on-air presentation at STAKES, customized survey forms just for STAKES, and tabulation of the survey information. That means you'll be on the air a minimum of *1,331* times between the first part of May and the end of September . . . *plus* you'll receive all the marketing information from a *$4,000 survey* that is included with this promotion. The price is one of the best parts. Your monthly investment is $13,010. Here's how it breaks down."

The rep turns the summary page (Exhibit 10.15) upside down and explains every line to the client and planner. She feels confident because she knows that her average unit rate looks good, that she has offered a lot of value-added services, and that she has studied her client's business from a marketing perspective.

Exhibit 10.15

Summary

Optimum Effective Advertising Campaign for May through September 199_ for STAKES ISUZU

May
- 4 Ladies' Nite Remote Broadcasts
- 152 Commercials

May 25–Labor Day
- Street Patrol STAKES ISUZU TROOPER Giveaway
- Promotion and Marketing Survey
- 675 Commercials

May–End of September
- Rotating OES schedules for STAKES ISUZU for every other week
- 352 Commercials

September
- 4 Ladies' Nite Remote Broadcasts
- 152 Commercials

That's a total of *1,331* times the STAKES name will be on the air from May through September!

Your monthly investment: $13,010

"Are there any questions?" (*At the moment, there are no questions.*)

"Good. The way to proceed with this plan is to fill out the paperwork and to decide which questions you'd like included on the marketing questionnaire. Would you like to use the survey form as it is now?"

Reality

We'd like to tell you that the client and the media planner said, "Yes, yes, yes, we love your ideas and we'll take the plan just as it is." In reality, they wanted to study the information, look at their budget, and listen to presentations from other stations. The bottom line is that our ABCD marketing rep did get a buy. Hope didn't get all of the $65,050 she was asking for but she got a *majority* of it. She went to a lot of trouble and work to get ready and make this presentation. Was it really necessary? You decide. Out of 26 stations, Hope's was one of 3 that got an advertising buy even though her station wasn't ranked first, second, or third in the demographic group the auto dealer was targeting. She also got an Isuzu Trooper for her station's promotional giveaway.

Did the media planner and client sit through the whole presentation? Yes. Were they bored with the amount of information presented to them? No. Hope focused mostly on increasing their business, selling their customers, and helping their salespeople.

11

A Niche Presentation for a Building Center

Insert Your Format

This building supply center presentation is telescoped to give an example of information that can be presented with *any* format. This kind of presentation can create a powerful niche marketing scenario for long-term business with your advertisers. As you'll see, your specific station's marketing information may be inserted after the cover page.

A sales rep, Michael, is about to make a presentation to the general manager of a building supply center that has four locations. A survey was conducted a week ago using the Niche Marketing Analysis: Part I. The rep's and advertiser's words are in quotes. The media rep begins:

> "Good morning, Crystal. I've really been looking forward to our meeting this morning because I've got some great information to share with you. I took the information you gave me during Part One of the Niche Marketing Analysis and studied it. I also contacted a marketing organization called the Radio Advertising Bureau, or as we call it, the RAB. The RAB is incredibly helpful to us at YYXX and to our advertisers. At RAB headquarters in Dallas, they have a staff of researchers who read about 400 trade magazines a month! You know the magazine you gave me to read, *Hardware Age*? Well, they also subscribe to it. They've been reading it from cover to cover for years! Even though a lot of people like you get the magazine and read it from time to time, your most important job, of course, is running this company. While you've been busy, the RAB researchers have been taking notes for you. I'd like to share some of the information we've uncovered about building supply customers.
>
> We've discovered what makes buyers go into a specific building center. And, we've discovered what they do and don't like when they get inside. But first, I'd like to share a little information with you about YYXX and how it fills a marketing niche in this area. Then, I'd like to show you how you can use that information to help you get more business. Take a moment to look at this."

155

Exhibit 11.1

(*Your station's letterhead*)

A Special Presentation
Developed for

CRYSTAL L. BRANHAM

General Manager
Alamo Building Center

Prepared by:

Michael K. Niles
Marketing Consultant
16 April 199_

INSERT YOUR STATION'S PRESENTATION
AFTER A COVER SHEET

The rep hands the general manager the cover page (Exhibit 11.1).

"During our initial interview you said that you spend about 35 percent to 40 percent of your advertising budget with newspapers. You told me that the newspaper has a broad appeal but it's wasteful because it has no specific target market. Well, since you use the newspaper so heavily, I'd like to share the results of a newspaper readership study that was conducted by a company called Starch INRA Hooper. This company has been measuring newspaper and magazine ad readership since 1923! I know that newspaper people rarely talk about *readership*. Often, the terms used are circulation and pass-along figures. This company conducts studies for newspapers but newspaper salespeople rarely hear about ad readership studies and they almost never see the results."

The sales rep, Michael, turns Exhibit 11.2 upside down so Crystal can see the page as he explains it.

"In this study, men and women were divided into separate groups and given newspapers to read. After they finished with the paper, one-on-one interviews were conducted. I'm going to share some of the results with you. As you can see, the two columns on your right are marked

Exhibit 11.2

Who Is Reading Newspaper Ads?

	Women Noted*	Women Read Most**	Men Noted*	Men Read Most**
Building/Hardware/Paint Stores*				
1 page or larger	37%	12%	43%	15%
3/4 to1 page	35	9	41	17
1/2 to 3/4 page	30	8	47	15
1/4 to 1/2 page	26	8	29	8
1/8 to 1/4 page	26	9	25	9
Under 1/8 page	13	6	16	7
All Ads	27	9	32	11

Conclusions

1. All newspaper readers are not ad readers. Circulation is no guarantee of readership.
2. Newspaper ads of smaller size deliver higher readership proportionately than larger ads.

***Noted**: Reader recalled seeing ad (did not have to recall content of ad).

****Read Most**: Reader recalled reading at least half of the ad.

***Studies were conducted by Starch INRA Hooper, a recognized leader in the measurement of newspaper and magazine ad readership. This report is derived from 6,400 personal interviews with male and female readers. The papers cover a cross-section of U.S. markets. The studies were conducted in all four seasons of the year in every region of the United States. Starch has been measuring newspaper and magazine ads since 1923, and the firm has conducted countless thousands of interviews with readers.

Women and *Men*. Under each heading you'll notice the words Noted and Read Most. *Noted* means that the reader recalled seeing the ad—they did not have to recall the content of the ad. *Read Most* refers to people who read at least 50 percent of the written copy in the ad."

As the rep explains the terms to the advertiser, he circles the words Noted and Read Most in each column. Michael wants the terms to stand out for his client and if anyone else in the company sees the information or the client looks at it later, this will help the reader focus on the most important aspects of the page. Michael circles the words *one-page-or-larger*.

"Let's look at the readers' responses when a one-page or larger ad was in the newspaper. First, let's look at respondents who *noted* or remembered seeing the ad when it was shown to them. Thirty-seven percent of women and forty-three percent of men did remember seeing the ad."

The sales rep circles the figures as he speaks. He then paraphrases so Crystal will have more of an opportunity to digest the information. Michael continues:

"That's a little more than ⅓ of women but less than ½ of men. Now, let's look at the more important column. It's more important because you obviously want people to *read* your ad. Only 12 percent of women and 15 percent of men reported reading *at least half* of an ad that was *one page or larger!* I was absolutely stunned when I encountered this study. I don't need to tell you that a full-page or larger ad in the newspaper is a sizable expense."

Crystal, let's look at ⅛- to ¼-page ads since you said that you often run ¼-page ads in the paper."

Michael then circles the words *1/8 to 1/4 page* and exclaims:

"When we look at women who *read most*, we see that only *9 percent* reported reading *at least half of the ad!* Only *9 percent* of men also reported reading at least half the ad! It's amazing. There certainly appears to be a relationship between ad size and readership. As ads get smaller, readership declines, *but not in proportion to size.* Notice that doubling the size of the ad *does not* double the readership, although it generally doubles the cost."

Mike emphasizes the point and concludes:

"What this study teaches us at YYXX is that smaller newspaper ads can be as effective as larger ads and they cost less. And, we promise that when we begin working together, we'll be happy to make print suggestions for any ad or promotion we present to you. We know that the most effective advertising is integrated. Our marketing strategies will be simple, easy, and encompass multimedia suggestions.

Do you have any questions or comments about anything we've discussed so far?"

Michael asks the question and pauses. Crystal relates a story about newspaper ad size and some pretty effective ads she has seen that occupied very little space. Michael is very attentive. He is enjoying her story as well as the presentation.

Mike turns the page (Exhibit 11.3) upside down and places it in front of Crystal, but keeps it within his reach also so he can use a colored marker to circle statistics.

"Crystal, here are some results from a survey taken by the Retail Hardware Research Foundation. This study looked at the ages of building supply home center customers over a 12-week period. As you can see, with the age ranges they've developed, the largest group of customers is

Exhibit 11.3

What are the ages of Building Supply Home Center customers?

18 to 34	31.6%
35 to 54	43.0%
55 and over	25.4%

How do customers shop?

One in party	64.5%
Two in party	30.0%
Three in party	6.5%

Source: Customer Shopping Patterns Study, Russel R. Mueller Retail Hardware Research Foundation, 199_.

---

between the ages of 35 to 54, then 18 to 34, and then 55 and over. Do you find that to be true with your stores?"

Mike waits for her response to the inquiry. Crystal tells Mike that she has a feel for the way her customer base is composed but admits that she really doesn't have any hard facts. Michael continues:

> "I'll have some suggestions for you for learning the age ranges—and a lot more about *your* customers—in just a few minutes. Right now, let's look at *how* customers shop. Since almost two-thirds of customers shop alone, it appears that they go into a building supply center ready to make purchasing decisions on their own. (*Mike has circled the 64.5% figure and the 30% figure.*) In other words, they aren't there to comparison shop and then take the information home with them. They're ready to buy. Does that make sense to you?"

Crystal agrees that it makes sense to her. Mike goes on and then asks for her opinion again.

> "This survey found that 30% of customers shop in pairs. Would it be accurate to say that these are probably couples who are interested in larger-ticket items?"

Crystal answers:

> "Yes. They're probably people that are planning additions or whole-room remodeling."

Mike asks:

> "Even though almost two-thirds of your customers shop alone, are they making smaller-ticket purchases?"

Crystal explains:

> "Yes. And, the one-third that shop as couples actually represent more profit because the vast majority of the time, they're buying the larger-ticket items."

Mike asks:

> "Even though your customer base is about two-thirds to one-third, does it make sense to you to run two separate ads? One that is directed to larger-ticket shoppers and one that is directed to the more frequent, lower-ticket shopper?"

Crystal responds:

> "Yes, it makes sense to me but the rotation of the ads should be 50–50 because it takes more of the one-party shoppers to equal the profit of the two-party shoppers."

Michael then proceeds to describe browsers before he puts the next page of his presentation on the desk in front of Crystal. He says:

> "Next, let's look at ways to entice shoppers into becoming browsers. While customers *pass through* an average of 12 or more departments, they stop and purchase in very few. This implies that few customers are browsers. They think they know what they want and they methodically get it—and usually it is one item in one department. Do you find this to be accurate?"

Crystal agrees that Mike is right. He places Exhibit 11.4 on Crystal's desk.

> "This situation is, of course, a challenge considering the need for more turns and greater volume. Wide, uncluttered aisles and well-signed departments can guide shoppers to the locations they're seeking. Easy-to-find departments, such as you have in your stores, show people just where they need to go to get what they want *(Crystal nods affirmatively)*. On the way there, attractive, attention-getting signage and displays can help slow customers down and entice them into seeing what other departments have to offer. *(Michael uses his colored marker to check off each item on his handout while he is speaking. This keeps Crystal's attention on what Mike is saying so she doesn't read ahead and miss his words.)*
>
> Feature BEST BUYS signage throughout the store to promote traffic flow and the purchase of nonsale items. *(Mike learned about BEST BUYS signage and its use during his NMA: I interivew.)* Shelf talkers and price stickers that

Exhibit 11.4

Entice Them into Becoming Browsers

Here's how . . .

    Wide, uncluttered aisles

    Easy-to-read signage

    Attention-getting signage and displays

    BEST BUYS throughout the store

    Combine merchandising with radio ads

    Easy-to-find shelf talkers and price stickers

Source: *Building Supply Home Centers* magazine.

---

are easy to read will help browsers become customers. These merchandising insights can be used to create effective ads for your stores. The BEST BUYS can be featured in your radio ads to tie-in with your store signage.

Is there anything else we should add to that list?"

Crystal thinks about what Mike has said as she looks at the list and then replies:

"No, nothing that I can think of right now."

Mike continues:

"Then, let's take a look at customers' pet peeves. You'll find this interesting."

To emphasize his promise, Michael puts the next page of his presentation (Exhibit 11.5) upside down on top of the last one in front of the manager.

"As you probably guessed, *not enough knowledgeable salespeople* is the overwhelming number one response. This information is also from the Building Supply Home Centers' Home Modernization Study. Yes, *items out of stock* was number two on the list. Then *indifference of salespeople*, followed by *poor how-to instructions* and *no point-of-purchase literature*. So far, is there anything from this survey that's surprising you?"

Crystal tells Mike that she's not surprised. She leans forward to continue reading the results. Mike asks:

"As you can see, *long lines at the checkout counters* and *no central information center* bring up the end of the list. Do you find that you have a lot less complaints about finding items since you installed your information desk?"

Exhibit 11.5

> **What are customers' pet peeves?**
>
> | | |
> |---|---|
> | Not enough knowledgeable salespeople | 70.1% |
> | Items out of stock | 49.2% |
> | Indifference of salespeople | 41.1% |
> | Poor how-to instructions | 35.6% |
> | No point-of-purchase literature | 28.1% |
> | Long lines at checkout stations | 20.1% |
> | No central information center | 16.4% |
>
> Source: Building Supply Home Centers' Home Modernization Study, May through September 1995.

Crystal relates several examples of customers complimenting employees about their information counter. Michael suggests that they should consider turning some of the negatives they've examined into positives. He recommends telling consumers in their radio ads that Alamo Building Center *has* knowledgeable salespeople by describing some of the backgrounds of a few of their sales associates. He proposes that they also cover some of the other pet peeves of building centers by mentioning the positive side of items on the list in future commercials. Mike promises to remind Crystal of this list on future visits, and she tells Mike that she likes the idea.

Mike places the next page (Exhibit 11.6) on top of the pile of sheets on the manager's desk. Then he continues:

> "I found some information about do-it-yourself customers from the National Retail Hardware Association. This study found that 74 percent of hardware store sales are to do-it-yourselfers and over 40 percent of lumber sales are to this group as well. During our last meeting when I interviewed you, you told me that these figures pretty much match your figures in your stores."

Crystal explains:

> "Within a few percentage points, our stores do match those. However, those figures can be flip-flopped in some building centers. It really just depends on which type of business the store decides to pursue and who are their competitors."

Michael suggests:

> "Let's look at the demographics for these do-it-yourselfers. (*Mike uses his marker to circle the figures for the male/female buyers.*) "I wasn't sur-

Exhibit 11.6

ALAMO BUILDING CENTER
D-I-Y Customers

Reaching D-I-Y Homeowners . . .

74% of hardware store sales are to D-I-Ys

40.6% of lumber sales are to D-I-Ys

Do-It-Yourself DEMOGRAPHICS for any remodeling activity in the past year:

Sex:	Male	61%	Age:	18 to 24	21%
	Female	39%		25 to 34	40%
				35 to 49	47%
				50+	33%

Source: National Retail Hardware Association Survey, 1994.

---

prised to see the breakdown between men and women customers because it's what you said you find in your stores. Females represent a growing segment of this do-it-yourself market. A few years ago, most people wouldn't have believed that almost 40 percent of a hardware store's customer base could be women buyers."

Crystal agrees:

"It's true. Even though my husband is a do-it-yourselfer, I buy a lot of our company's products and enjoy using them for projects around our house."

Mike continues:

"I looked at the age ranges of these customers and tried to understand their needs. The 35- to 49-year-olds represent the largest group of buyers. I figured that this group represents young families. They probably try to do a lot of home repairs and improvements themselves. Studies say that the majority have young families and that they are trying to stretch their money as far as they can. Do you think that's accurate?"

Crystal agrees once again:

"That's been my experience. I fit in that group, too."

Michael goes on:

"The 25- to 34-year-olds represent your next largest segment of shoppers. This portion of your customers are most likely first-time homeowners. They've invested a lot in just getting set up in a home of their own and are adjusting to the hidden costs of ownership such as real estate taxes and homeowner insurance. As a whole, they really need to do their own repairs or fix-up projects. Do you agree with that scenario? (*Mike questions, and the manager answers, "I do."*)

The 50-years and older crowd have been in their homes longer, their families are ready to leave the nest or are grown and gone. This segment of buyers doesn't need as many home repairs because they've been making repairs for a number of years. For the most part, they aren't doing major home improvements or building additions themselves. They can afford to hire professionals to do the work. Does that sound logical to you?"

Michael pauses and Crystal responds:

"Yes, my parents fit into this category and you've just described them—except for my younger brother who keeps moving back in with them!"

Mike responds:

"Oh well, I went through my *terrible twos* the second time—some people say that's the twenties. And, I moved back with my parents for a short while. I understand it well!

The last group is the 18- to 24-year-olds. I can see why they represent the smallest segment of customers. The majority of this group doesn't own homes yet. Their largest purchases at this point are stereos and cars."

Crystal adds:

"That's my brother!"

Mike inquires:

"Are there any questions or comments on anything we've discussed so far?"

Crystal answers no and Mike goes on:

"Then, let's look at some marketing and advertising suggestions I've developed for Alamo based on the interview we had last week and based on the findings I uncovered during my sojourn through our research files and your industry's trade journals."

Three Choices

Michael has three plans for Alamo's general manager to consider.

Plan One. The first plan is for 26 weeks. It is based on the frequency Crystal told Mike she has used on other stations.

Plan Two. The second plan is for 26 weeks based on a higher frequency of ads; it also contains a proposal to conduct one-to-one research with Alamo's customers and browsers. The plan offers to deliver survey results at the end of 13 weeks and at the end of 26 weeks. Alamo Building Center will supply prizes for a monthly drawing to reward consumers for filling out entry/survey forms. The price of the survey has been built into the package.

Plan Three. The third plan is for 52 weeks. The frequency is similar to the level suggested in the first plan, but it contains more services. The customer/browser survey is a part of this plan, but Mike offers to deliver *monthly* reports from survey results. This will allow Crystal and Mike to make adjustments to their marketing strategies more often if necessary. This plan also will give the manager a more complete picture of her customers' wants and needs because it is spread over a longer period of time. Mike explains that the most reliable and insightful findings are a result of gathering bits of information over a year rather than a short survey, which more likely will focus on customer needs for that season.

Additional Services

After Michael presents his three plans, he tells Crystal about their focus group studies performed for other business leaders in the area. He gives her a separate sheet of paper that tells more about focus group research. The page shows how easy it is to work with YYXX on this type of project. The sheet also gives prices for single or multiple focus group interviews. As you would expect, when an advertiser purchases more sessions, the investment for each session declines. Michael doesn't expect Crystal to purchase a focus group study at this meeting. He is laying the groundwork, however, for future business. He wants to give her an opportunity to begin thinking of Alamo as a company that invests in these services from his radio station.

Closing Takes Time

Crystal and Mike discuss the customized survey research, focus groups, and possible marketing strategies for Alamo for the next 30 minutes. At the end of that time, the manager agrees to sign a contract for Plan Two. Mike is careful to describe the services that are included with the advertising frequency. He then makes another appointment to meet with Crystal so Michael can make plans to conduct the Niche Marketing Analysis: Part II survey and discuss questions Crystal wants to put on her customer survey form.

Mike, The Futurist

Michael knows that he has 26 weeks to deliver the best service Alamo has ever received from any media. He feels very confident that 23 weeks from now he will be leaving Crystal's office with a 52-week contract that will include future research projects the station will be conducting. Life is good.

12

Tabulating Survey Results

The First Step

Congratulations. You've sold your radio station's services for a customized consumer survey to Murray Auto Supply. You've been trying to upgrade Jim Murray for a long time. Finally, you saw a light appear in his eyes when you asked him "What would you like to know about your customers?" After he signed your 10-week contract, you produced an ad for him and began airing it. Next, you both focused on developing a registration/survey form. Your station provided the entry box, registration forms, and posters for Jim. The posters told people that they could win a $50 gift certificate for completing the form. The survey will be conducted for eight weeks. At the end of that time—allowing at least two weeks for tabulation and report writing—you will deliver the results to Jim. Within about a week, you gave birth to the form in Exhibit 12.1.

The Survey Is Completed—Help!

You have diligently collected completed entry forms each week of the Murray survey. You aren't going to report statistics week by week because you sold Murray Auto Supply a study that is to report findings for the total 8-week period only. If Jim had opted for a breakdown of consumers week by week, he would have made a larger investment. Jim has been able, however, to chart his store's traffic flow by number of entries received weekly. The number of entries is *not* a representation of how many customers were actually in stores because not everyone fills out an entry form—though some people do fill out several forms in the hopes of increasing their odds of winning. And, not all businesses give a reward for filling out survey forms; however, a prize does increase the number of responses your client will receive.

Exhibit 12.1

Murray Auto Supply Questionnaire

OFFICIAL REGISTRATION *(Radio Station Logo)*

PLEASE FILL OUT THIS FORM AND DEPOSIT HERE. ENTER AS OFTEN AS YOU VISIT THIS LOCATION. YOU MUST BE 18 TO ENTER.

Murray Auto Supply visited:
☐ Glendale ☐ Mesa ☐ Scottsdale ☐ Peoria

What kind of maintenance do you perform for your car (e.g., oil change, tune up)? _____

Rate, in order of importance, the factors that determine where you buy auto parts:

____ Location ____ Parts availability ____ Selection ____ Hours or days open

____ Knowledge of counterperson ____ Price ____ Other: _____

What do you like about Murray Auto Supply? _____

What can we do to improve our stores for you? _____

Glendale: _____

Mesa: _____

Scottsdale: _____

Peoria: _____

My age range: ☐ 18–24 ☐ 25–34 ☐ 35–44 ☐ 45–54 ☐ 55–64
☐ 65–74 ☐ 75+

Name: _____ Phone: _____

Street address: _____

City: _____ State: _____ Zip: _____

Making Sense Step-by-Step

Counting

You have a very large bag of survey forms. You may have an intern or sales assistant who will tabulate responses for you. Because you already know the questions on Jim's survey form, you can have the tabulation forms printed (Exhibit 12.2). You are organized, of course, and each week you have *counted* the number of entries before you added them to the previous week's forms. To avoid feeling overwhelmed and adding a lot of stress to your life, you have *tabulated* your entries each week also. When all the entries have been counted, you can now define your total

Exhibit 12.2

Murray Auto Supply Tabulation Form

Week #: _____   *(Use this line if you are furnishing week-by-week statistics.)*
Total entries: _____

Murray Auto Supply visited:
Glendale: _____ Total: _____
Mesa: _____ Total: _____
Scottsdale: _____ Total: _____
Peoria: _____ Total: _____

What kind of maintenance do you perform for your car?
All: _____ Total: _____
Oil change: _____ Total: _____
Tune up: _____ Total: _____
Minor repairs: _____ Total: _____
Overhaul: _____ Total: _____
Brakes: _____ Total: _____
Other: _____ Total: _____

Rate in order of importance:

	1	2	3	4	5	6	7
Location:	☐	☐	☐	☐	☐	☐	☐
Parts availability:	☐	☐	☐	☐	☐	☐	☐
Selection:	☐	☐	☐	☐	☐	☐	☐
Hours/days open:	☐	☐	☐	☐	☐	☐	☐
Price:	☐	☐	☐	☐	☐	☐	☐
Knowledge of counterperson:	☐	☐	☐	☐	☐	☐	☐
All of the above:	☐	☐	☐	☐	☐	☐	☐
Quality:	☐	☐	☐	☐	☐	☐	☐
Other: _____	☐	☐	☐	☐	☐	☐	☐
_____	☐	☐	☐	☐	☐	☐	☐

Totals for above categories: _____
Location: _____
Parts availability: _____
Selection: _____
Hours or days open: _____
Price: _____
Knowledge of counterperson: _____
All of above: _____

Quality: _____
Other: _____

What do you like about Murray Auto Supply?*
_____
_____
_____

What can we do to improve our stores for you?†
Glendale: _____	Total: _____
Mesa: _____	Total: _____
Scottsdale: _____	Total: _____
Peoria: _____	Total: _____

Age range:
18–24: _____	Total: _____
25–34: _____	Total: _____
35–44: _____	Total: _____
45–54: _____	Total: _____
55–64: _____	Total: _____
65–74: _____	Total: _____
75+: _____	Total: _____

Zip code:‡
Glendale: _____	Total: _____
Mesa: _____	Total: _____
Scottsdale: _____	Total: _____
Peoria: _____	Total: _____
Phoenix: _____	Total: _____
Paradise Valley: _____	Total: _____
Tempe: _____	Total: _____
Gilbert: _____	Total: _____
Other (out of state/area): _____	Total: _____

*This is an open-ended question, so expect a variety of responses. You will encounter, however, many similar responses. This is where research becomes less scientific and more intuitive. You must decide if two similar words should be combined under one category. Take heart, you will see patterns begin to emerge to help guide you.

**You're again faced with making omnipotent decisions. If your client has multiple locations—and they purchase a breakdown for each location—you'll often find that you will receive very different responses for each store.

†This category—depending on whether your advertiser *purchases* an in-depth survey—can be broken down by *street*. This information can be a gold mine for advertisers who use direct-mail or telemarketing campaigns. Politicians are also good candidates for this intelligence.

"universe" for the study. If you have 600 nonduplicated entries, that is the total universe.

Location Visited

An advertiser can learn a lot about customers' perceptions of stores based on a survey like this one. For instance, you can categorize all of the Glendale entries by the kinds of service customers perform on their vehicles. Your client may need to stock more of particular kinds of products to fill those maintenance needs in the Glendale store. As a result of your survey, Jim may find that he would be wise to stock a different quantity of the same products in another location. This data alone could save the owner of an auto parts store—or any store—a lot of money.

Kinds of Maintenance

Data from your study could disclose that a specific kind of maintenance is performed mostly by two age ranges of customers. You may find, for example, that brake replacement is carried out mostly by people between the ages of 18 and 34. Your client can use this intelligence to target consumers on radio stations with formats that attract 18–34-year-olds. "Your format doesn't do well with that demo," you say.

Examine the figures for your station's listeners. Study the responses for your demographic. How much can you learn about them from this study? In what ways can you help your client target *your* audience? Look at your survey. What is important to consumers in *your* demo? Tell them in ads that your client's stores have the products and/or services that are important to them. Your client and you will be more successful. Help your clients reach consumers who are not in your station's demo. Contrary to popular belief, you won't lose business. You will be gaining trust. You'll be building a valuable partnership. Whatever you give others is also a gift to yourself. Remember, consultants advise as well as sell.

What Is Important to Consumers

Finding out what consumers want and then giving it to them is the golden key to the success of any business—including yours. If respondents tell you that parts availability is the most important reason they shop your client's store, include that in your ads. If consumers between the ages of 25 and 44 tell you that knowledgeable counterpeople are important, use that information in your advertising campaigns. If you find that 25–44-year-old consumers who want knowledgeable counterpeople exclusively visited one location, your client has just been given a red warning flag about possible personnel problems in other stores. There are a multitude of ways you can help your client with the data you uncover. Begin to look at your client's business from the perspective of being his silent partner.

Duplication

When you conduct consumer surveys where prizes are awarded, you'll usually find that a few individuals from each location will stuff the registration box with multiple entries. Even if you are dealing with large numbers of entries, it's possible to isolate the duplicates. Count duplicates as *one* entry when you're compiling statistics. Tell your advertisers before the study begins to expect enthusiastic browsers and customers to register often. We're not trying to add to your workload, but we suggest that you tell your clients to encourage their patrons to enter as often as they wish. The customer sees this as an opportunity to *win* something—not as a survey. *And*, don't forget to count the winning entry.

Percentages

Now that you've tabulated all of the categories from your survey forms, you need to change your numbers into percentages. If you have a software program you're using, it may perform that task for you. If you're compiling the information by hand, you will find that the process is easy but time-consuming. When you are changing numbers into a percentage, always remember to divide the smaller number by the bigger number. In other words, take the total number of 55–64-year-olds who responded to your survey and divide that number by the total number of entries. Now, you will know that __% of respondents (e.g., 10%) were between the ages of 55 and 64. If you're trying to define the composition of your client's customer base, you've taken your first step. Numbers will give you an idea about the end results of your survey. Percentages, however, will be like seeing a fuzzy picture suddenly come into sharp focus. Bar graphs and pie charts will also help you illustrate the "Customer Story" for your client.

Unlimited Ideas

If you're lucky enough to have a software program like Borland's Quatro Pro for Windows, you'll soon discover an enormous number of ways to draw information from your statistics. If you're compiling reports by hand, you will still become excited by the variety of ways you can look at an advertiser's customers—and noncustomers. You'll be able to develop an easy-to-see picture of consumers' physical movements. For example, you'll see where customers travel from to get to your client's locations. You'll discover if a younger age range of customers has a different view of how to improve your client's store versus an older age range of customers' views. If your advertiser wants to cater to one age range more than another, studies like these can give the store an insight into what is important to that age group. This is truly an example of one-to-one marketing—from the "mouth of the customer" to the business owner. An incredibly successful niche can be carved from this information—it has enormous value.

Presenting Your Findings

Now that you've compiled your results and you're bursting with ideas to share with your client, Chapter Thirteen provides an example of how to present your findings to your advertiser to help you get a renewal.

13

Presenting Survey Results

Searching for Niches

Bicycle shops may not be among radio's top-ten advertising categories but the surveys one radio station conducted for a very profitable bike shop yielded a long-term client who steadily increased his advertising buys. The advertiser's marketing decisions were based, in part, on information from customers through the use of ongoing surveys like the one in Exhibit 13.1.

The market is a mid-Atlantic resort area in the United States. This client had participated in a one-to-one marketing survey, which was combined with an advertising promotion. The report from the survey, which the station called *A Marketing Blueprint*, was used to garner a renewal from the client. After reviewing this scenario, you should be able to compose your own reports based on clients' one-to-one marketing surveys of their customers, leading to long-term marketing partnerships.

Exhibit 13.1

Niche Marketing Questionnaire for Beach Bikes & Blades

REGISTRATION (*Your Station's Logo*)

PLEASE COMPLETE THIS FORM AND DEPOSIT HERE.
ENTER AS OFTEN AS YOU VISIT THIS STORE.
YOU MUST BE 18 OR OLDER TO ENTER.

Do you repair your own bicycles? ☐ Yes ☐ No

Type of bike owned now: ☐ Beach cruiser ☐ 10-speed
☐ Mountain bike ☐ Other

How many bikes do you own?
☐ 0 ☐ 1 ☐ 2 ☐ 3 ☐ 4 ☐ 5 or more

Are you ☐ Vacationing ☐ Resident ☐ On business
☐ Other _____

Are you planning to buy a bike in the next 6 months?
☐ Yes ☐ No ☐ Don't know

If yes, how much do you expect to spend? _____

On average, how far do you cycle each week?
☐ up to 5 miles ☐ 6–10 miles ☐ + 10 miles

Other than this promotion, how did you find out about us? _____
_____

Age range:
☐ 18–24 ☐ 25–34 ☐ 35–44 ☐ 45+

Name: _____ Phone: _____
Street address: _____
City: _____ State: _____ Zip: _____

1. DO YOU REPAIR YOUR OWN BICYCLES?

Yes - - - - - - - **63%**
No - - - - - - - - **29%**
No reply - - - - - **8%**

Since a major percentage of the respondents (63%) say that they repair their own bikes, an ad about your selection of parts for "do-it-yourself" bike repair would be in order.

Almost *one third* of respondents report that they don't repair their own bikes. That percentage warrants an ad featuring your bike-repair facilities. You may choose to give this ad a lower rotation schedule than ads that address the majority of buyers.

Page 1

A Marketing Blueprint

Cover Sheet

Beach Bikes & Blades – Coast 93.1

M A R K E T I N G   S U R V E Y

Conducted February 21 to March 17, 199_

In

Resort Area, USA

Page 2

2. TYPE OF BIKE OWNED:

Mountain bike - - - - - - 61%
10-speed - - - - - - - - - 26%
Beach cruiser - - - - - - 21%
Write in: 20" - - - - - - - 5%

The results indicate that separate ads for each bike, rotated according to sales volume, would give the best return on your advertising investment. For example, the mountain bike ad may run two times while the other ads run once each.

Page 3

3. HOW MANY BIKES DO YOU OWN?

2 bikes - - - - - 24%
3 - - - - - - - - - 16%
1 - - - - - - - - - 13%
5 - - - - - - - - - 11%
+5 - - - - - - - - 11%
4 - - - - - - - - - 3%
0 - - - - - - - - - 3%

If the consumers who own five bikes or more are buying them from you, you have an excellent share-of-customer rather than share-of-market. If not, you may want to offer package pricing for multiple purchases.

Page 4

4. ARE YOU:

```
Resident- - - - - - - - 76%
Vacationing - - - - - - 11%
On business - - - - - 11%
Second-home owner -  2%
Other - - - - - - - - -  0%
```

Because the survey was conducted in February and March, it's no surprise that the overwhelming majority of respondents were locals. The fact that 11% of the people surveyed were on vacation seems to be an indication of a fairly healthy economy at this time.

The county Chamber of Commerce has reported an increase of 23% in retail sales from this time last year to now. That figure substantiates your findings.

Page 5

5. ARE YOU PLANNING TO BUY A BIKE IN THE NEXT 6 MONTHS?

```
Yes - - - - - - - 45%
No - - - - - - -  42%
Don't Know - -   11%
No reply - - - -   2%
```

According to your respondents, 56% may be in the market for a bicycle in the next 6 months. This indicates that a majority of people who browse in your store are buyers—or could be—with help from your sales staff.

Because we don't know why 42% do not plan to buy a bike in the next 6 months, it isn't possible to determine if they are in the store for parts, out of curiosity, or they didn't want to answer "yes" and take the chance that there would be follow-up from the store through direct mail or a phone call.

6. HOW MUCH DO YOU EXPECT TO SPEND?

```
$   400 - - - - - - - -  8%
    500 - - - - - - - -  5%
  1,000 - - - - - - - -  5%
  1,500 - - - - - - - -  5%
    700 - - - - - - - -  3%
    250 - - - - - - - -  3%
    225 - - - - - - - -  3%
    200 - - - - - - - -  3%
    100 - - - - - - - -  3%
Don't know - - - - - -   9%
No reply - - - - - - -  53%
```

The survey results show that the largest percentage of your customer traffic expects to pay about $400 for the next bicycle he or she purchases. Do you sell a lot of bikes in this price range? If you use price-and-item advertising, featuring a bike in this price range would seem to be appropriate.

Because 53% did not reply, this may indicate that consumers just don't know how much they may have to pay or how much they will be spending.

7. ON AVERAGE, HOW FAR DO YOU CYCLE EACH WEEK?

```
+10 miles - - - -  37%
5-10 - - - - - - - 29%
1-5 - - - - - - -  26%
No reply - - - - -  8%
```

Notice that 37% of your customer traffic say that they ride over 10 miles per week. To target this group, including phrases in your ads about "serious cyclists" should be beneficial.

Recreational bikers represent 26% of your respondents. A separate campaign for this segment of your customer base makes sense.

Presenting Survey Results **179**

Page 8

8. OTHER THAN THIS PROMOTION, HOW DID YOU FIND OUT ABOUT US?

```
Referral  - - - - - - - - 26%
No reply  - - - - - - - - 25%
Walk-in-  - - - - - - - - 21%
Newspaper Ad  - - - - -  11%
Been in before - - - - -   8%
Boys & Girls Surf Shop -  3%
Andy  - - - - - - - - - -  3%
Radio Ad  - - - - - - - -  3%
```

We translate "referral" as word-of-mouth. We seldom know, however, how word-of-mouth is generated.

In other marketing situations, we find that a majority of adult consumers have no idea how they heard about a particular business.

Page 9

9. AGE RANGE:

```
18–24 - - - - - 42%
25–34 - - - - - 29%
35–44 - - - - - 13%
45+   - - - - - 13%
No reply - - -   3%
```

In this market, the most effective advertising medium to reach 18–24-year-olds is radio. We recommend Rock 105 (this in not the station writing the report) to reach this group.

According to this study, 25–44-year-olds represent 42% of your customer base. Coast 93.1 and Kissin' 92 are the most effective media for this demographic.

For the 45+ group—who represent 13% of your customers who responded—WXXX, Bubba 101.5, and newspapers make sense.

With the media choices available in this market, our overall recommendation for reaching the majority of your very active, 18–44-year-old customer base is radio.

(*Authors' Note*: There is no local television. Cable penetration is not very effective because tourists account for approximately 75% of most businesses' revenues. This market is filled with a majority of people pursuing very active lifestyles. Cable television must compete with beachgoers, deep-sea fishing, surfing, and numerous other attractions.)

Your Customers Respond

Suggested ads (based on consumer responses to survey):

1. An ad featuring **your selection of parts for do-it-yourself bike repair.**
   (63% report repairing their own bikes.)
2. An ad for **Mountain Bikes.**
   (61% of respondents own this kind of bike)
3. Ads for **10-speed bikes** and an ad for **Beach Cruisers** rotated with ad for Mountain Bikes.
4. Ads featuring **package pricing for multiple bike purchases**
   (24% of your store traffic own two bikes).
5. An ad featuring your **bikes in the $400 price range.**
   (Median price range)
6. Ads targeting **serious cyclists.**
   (37% of store traffic rides 10+ miles per week)

Continue Learning What Customers Want

Based on information gathered from the 3-week survey of mostly local residents, Coast 93.1 suggests the following advertising plan as a continuation of the momentum you've developed during February and March.

Your specific advertising proposal here.

Looking Forward

Continually gathering small bits of information from consumers over a long period of time will result in the most accurate picture of what buyers want. Encourage your advertisers to constantly be collecting intelligence from their customers and noncustomers. These bits of information will help businesses assemble the wants-of-consumers jigsaw pieces into a cohesive "photograph."

14

Introducing Niche Marketing Services to Advertisers

A Niche Marketing Presentation

The how-tos of welcoming your advertisers to the new era of niche marketing are covered in this chapter. The presentation here teaches your clients and prospects how to collaborate with *their* customers to increase share-of-customer rather than share-of-market. This method also shows *you* how to increase share-of-advertiser. You'll discover how one salesperson shows a client how to begin a dialogue with customers to unearth heavy users. You'll also learn how to sell customized local research to advertisers.

The headline on the title page should catch the attention of the recipient and give him a reason to continue listening. Beth hands the cover page (Exhibit 14.1) to her client. She silently waits for Marshall to read the introduction. Then, Beth begins:

> "Please take a moment, Marshall, to look at my introduction. I'm here this morning to tell you about a death. It's not a sad story but it's an important one.
>
> The mass-marketing era as we've known it is over. Mass marketing is dead. May it rest in peace. The old systems of mass production, mass media, and mass marketing are being replaced by a totally new one-on-one economic system. This new system is changing the way you and I will be conducting business in the future. Before we examine this new economic system though, let's take a brief look at some of the ways we've reached customers in the past.
>
> The first 100 years of this country's existence, we had an agrarian-based economy. People were scattered in mostly rural areas. News—in the form of letters—often arrived in very remote areas *months* after an event had taken place. Our country could elect a new president and some people wouldn't know about it for several months! *(Beth exclaimed.)*

Introducing Niche Marketing Services to Advertisers   **183**

Exhibit 14.1

What's Really Going on in the Minds of Your Customers?

A Special Presentation

for

Marshall Simon

President

GWYNN'S of Mt. Pleasant

Prepared by:
Beth McClellan
Marketing Consultant

The second 100 years our country evolved into an industrial-based economy. People started to gather where they could find jobs—in factories. Cities grew around the factories. The majority of our workforce became centralized. Getting your message to the masses was possible. Manufacturers and retailers adopted an *offer-a-product-and-they-will-buy-it* philosophy.

Then, during the next 100 years—during our lifetime—changes in our economy escalated. As you know, we now have an information-based economy. With the help of the computer, we started studying ourselves more closely. Researchers found that markets—previously thought of as an accumulation of people with very similar wants and needs—didn't exist anymore.

Instead, they discovered that we are really a lot of different kinds of people with decidedly diverse tastes and needs. The four-person, nuclear family with dad, mom, two kids, a cat, and a dog is *not* in the majority. Today, because of career moms, unrelated people living under the same roof, latchkey kids, and an eclectic grouping of people with strong ethnic ties, our markets are fragmented. Methods for reaching these groups have changed also. This page covers what I just told you."

Beth pauses and hands Marshall Exhibit 14.2.

184 *Presentations and Niche Copywriting*

Exhibit 14.2

> # R.I.P
> ## Mass Marketing Is Dead

The old system of **mass production ... mass media ... and mass marketing** is being replaced by a totally new one-to-one economic system.

Offer-a-Product-and-They-Will-Come Philosophy

Reaching Customers

The First 100 Years ... An Agrarian-based Economy
People were scattered in mostly rural areas. News often arrived MONTHS after an event took place.

The Second 100 Years ... An Industrial-based Economy
People gathered where the jobs were found ...
in factories in cities. The workforce was centralized.
Getting your message to the masses was possible.

The Next 100 Years ... An Information-based Economy
Markets splintered ... because of single-person households, unrelated people sharing living quarters, career moms, latchkey kids, senior citizens, single-parent families, and an eclectic collection of ethnic groups. Our methods for reaching these groups have changed also.

"Today, only about 8 percent of businesses are using niche marketing, but marketers have predicted that by the year 2000, *80 percent* of businesses will be using these techniques. How will the niche marketing era affect Gwynn's? What do you do in the retail clothing business to survive—and grow?"

Beth asks these questions and then hands the next sheet (Exhibit 14.3) to Marshall. She waits patiently while he looks at it. Then, once he has put it down on his desk and reestablishes eye contact, she continues.

"Let's look at the difference between mass marketing methods of the past and niche marketing techniques of the future. A shoe store runs an ad. 'We're having a sale. 50% off all shoes. We have shoes for men, women, teens, kids, and senior citizens,' the ad says. That's an example of mass marketing. We hear and see ads like that all the time. Don't you agree?"

Beth pauses again after asking the question. Marshall agrees.

"That's making one offer to multiple segments. With niche marketing, however, we need to take a closer look at each of the shoe store's customer types."

Beth explains as she takes the next sheet (Exhibit 14.4) out and turns it upside down and places it in front of Marshall on his desk. She leans over the desk to elaborate. As Beth explains, she circles the ad with her marker.

"This is a typical shoe store mass-marketing appeal. Now, let's look at this offer from a niche marketing perspective. Each group of buyers—whether it's teens, kids, men, whoever—has different reasons for buying shoes at that store. For example, women who go into Paul's Shoe Store would go in to buy shoes for a different reason than senior citizens or teens or any of the other segments. To effectively attract each of the five segments, we may be looking at five different ad campaigns."

Beth clarifies by putting a check beside each numbered segment as she speaks. She returns to her seat.

"Let's look at some of the ways the mass marketing giants are dealing with fragmenting markets. First, let's look at a very successful mass merchandiser—Coca-Cola. Until a few decades ago, if you wanted to buy a Coke, your choices were simple. There was no Diet Coke or Caffeine Free Coke. But what has happened to Coca-Cola? When you go into a food store today, you will find a wall of different kinds of Coke. There is Coke Classic, Diet Coke, Caffeine Free regular Coke, and Caffeine Free Diet Coke. Does anybody want a Coke? Today, you have to give an elaborate description of *which* kind of Coke you feel like drinking. This is a result of the Coca-Cola company's attempt to reach out to many differ-

186 *Presentations and Niche Copywriting*

Exhibit 14.3

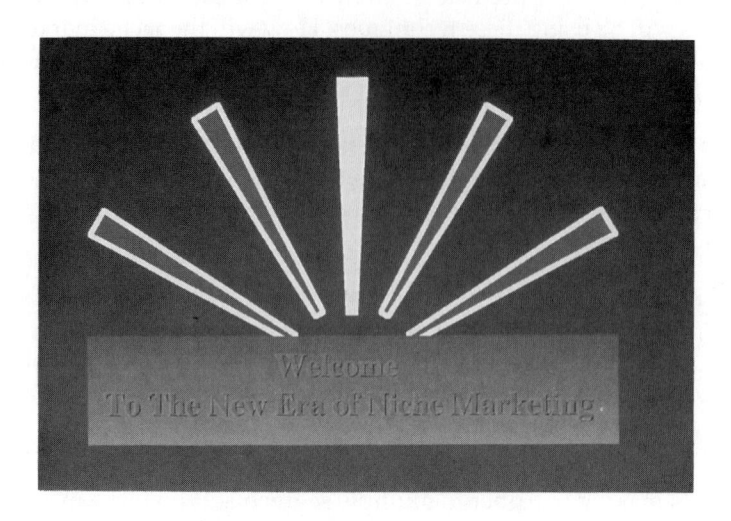

Welcome To The New Era of Niche Marketing

Market fragmentation . . . happening all over the world leads us to the new era of Niche Marketing.

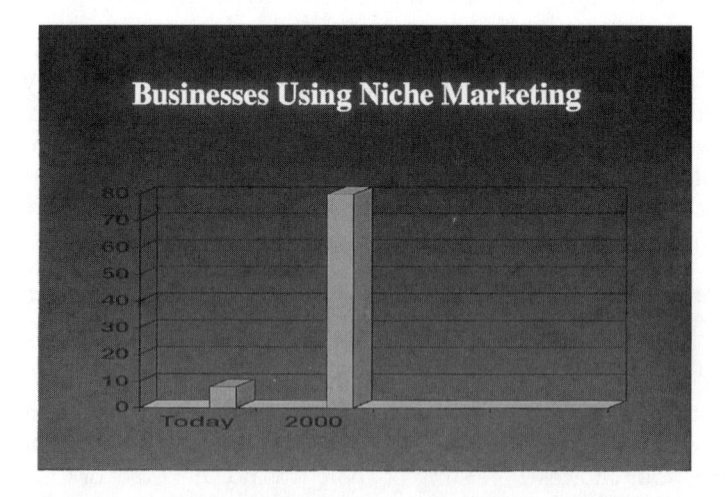

Businesses Using Niche Marketing

Today, only 8% of businesses are using niche marketing, but it's been predicted that by the year 2000, 80% of businesses will be using it.

Exhibit 14.4

With Mass Marketing
. . . you appeal to multiple segments of your customer base with one offering.

Paul's Shoes
The Family Shoe Store

Clearance

50% Off ALL Shoes!

MENS WOMENS TEENS
SENIORS CHILDREN

555 Main Street   Open Mon-Fri 9-5
Big City, Big State   Sat   9-5
555-4321   Sun closed

Shoe Store . . . 50% Off ALL Shoes!
Shoes for Men, Women, Teens, Seniors, Children

With Niche Marketing
. . . you split the shoe store's mass market into 5 segments:

Shoe Store Marketing Niches
1. CHILDREN'S SHOES
2. WOMEN'S SHOES
3. MEN'S SHOES
4. TEENS' SHOES
5. SENIORS' SHOES

Now . . . take a close look at **EACH** one of those 5 different segments.

ent kinds of consumers who have dissimilar tastes and needs. Coca-Cola has recognized the new era of niche marketing.

The same type of niche marketing techniques are being applied by Reebok. Think of all the choices for athletic shoes. Remember when the only choice was high tops or low cuts? Today, Crest toothpaste offers *twelve* different kinds of Crest. American Express has joined the niche marketing parade, too. You can now possess the original green American Express card, the Corporate Card, the Gold Card, and the Platinum Card. The niche marketing train gets longer as we speak. This page shows what we've just discussed."

Beth tells Marshall all this and then she hands Marshall the next sheet (Exhibit 14.5). Beth waits for Marshall to read the page.

Exhibit 14.5

Consider some of the niche marketing leaders . . .

COCA-COLA

REEBOK

CREST

AMERICAN EXPRESS

"What's the first step? How do you look for niches you can serve profitably? Well, the first step is to look for your heavy users. We want to study customers who are most likely to buy more of your inventory, more often, than the majority of your customers. Then, we need to find out what they absolutely *love* about your store—not just what they *like*. Let me give you an example of an automobile dealer's heavy users."

After Beth suggests this, she hands the next sheet (Exhibit 14.6) to Marshall.

"On one side of the scale, there's an auto dealer's customer. This customer buys an automobile from the dealership. He's pretty happy with the car. Within a short period of time, he buys a second automobile (while keeping the first one). Again, he's thrilled with the product and also the service. Not long after that, he buys a third automobile. This consumer is an example of that dealership's heavy users. The owner of the dealership wants to learn more about this very satisfied customer. She wants to know what motivates him to buy—time after time. As marketing partners with the dealership owner, we want to talk with her heavy users. Once we understand why they're thrilled with that business, we can use the information in her radio ads to bring in other potential heavy users. Business will flourish. Are there any questions or comments about anything we've discussed so far?"

Exhibit 14.6

What's the first step?
How do you begin a dialogue with your customers?

Look for your heavy users:

An automobile dealer's heavy users

Marshall looks over the papers after hearing Beth's question. He doesn't have any questions so far, he says. Beth continues:

> "Here's a list of some of the services our radio station is now offering to clients. Let's go over some of these services so that I can explain what that might mean to Gwynn's."

She gets up and goes behind the desk to stand next to Marshall. Beth places Exhibit 14.7 on his desk. She proceeds to explain to Marshall the services listed on the sheet:

> "The first type of survey on the list is an intercept survey. These may be better known as person-on-the-street interviews. Of course, they may be conducted inside or outside, but basically interviewers stop people and ask them a series of questions. The respondents may be consumers who may or may not have been in your store. This kind of survey is often useful when you want to learn about your store's name recognition in the market or about top-of-mind awareness of Gwynn's.
>
> Contest surveys usually occur when we bring the entry/survey forms to your store. People enter here by filling out the forms. We know that these consumers have actually been inside your store. We can ask more specific questions about your store's inventory, layout, and service than the person-on-the-street interviews.

Exhibit 14.7

**How to Get Business From Your Competition
With Customized Local Research**

1. Intercept Surveys
2. Contest Surveys
3. Focus Group Surveys
4. Telephone Coincidentals
5. Exit/Parking Lot Surveys
6. Direct-Mail Surveys
7. Customer Surveys
8. Consumer Surveys

You'll sell more by surveying your customers.

Of any type of survey, the focus group studies we perform can give you the most in-depth information about Gwynn's. The intelligence, however, is coming from a much smaller number of people. Many useful insights, which can help Gwynn's plan marketing and advertising strategies, can be gained from these sessions. These insights may also help your buyers make purchases more tailored to the wishes of local consumers. You may learn ways to improve service. Focus group surveys have often uncovered image problems of which business owners have been unaware. Have you ever commissioned a focus group study?"

Marshall's answer to Beth's question is that he hasn't. Then, Beth goes on:

"Telephone coincidentals may be conducted in one day or may be extended over several days. Usually, you'll have results sooner with this category of survey. You can easily target the geographic location you survey. This is helpful if you wish to know how consumers in a particular part of the market feel about your store.

Exit/parking lot surveys can target buyers or nonbuyers based on whether they are bringing merchandise out of the store with them. Interviewers also qualify respondents by asking if they have just made a purchase. These studies may focus on why people *don't* buy. That intelligence may be as valuable to you as why people buy.

Direct-mail surveys, like telephone coincidentals, can target a specific community or section of the market. Questionnaires can be fairly lengthy if they are easy to read and the respondent can answer almost all questions by simply checking boxes. Since this survey requires that respondents motivate themselves to participate, an opportunity to receive a reward is included. This reward may be a chance to win a gift certificate from your store." *(Beth pauses and then continues.)*

A customer survey can disclose what customers say motivates them to buy. You may choose to target only your heavy users or just your light users. Each group may relate unknown information you need. These surveys may be conducted in-person, over the phone, or through the mail. Have you ever conducted a survey of your customers?"

Marshall says that he hasn't purchased any customer studies. Beth explains more:

"Consumer surveys usually do not include your customers. Like the intercept survey, they tend to be more general in nature. For example, you may choose to gain knowledge about your store's image and about consumers' awareness level of your business. Unlike intercept surveys, consumer surveys may be conducted through telephone interviews, in the mail, or face-to-face. Do you have any questions or comments about any of the types of surveys we've discussed?"

Beth stops after she asks the question. Marshall tells her that he would like a little more information about focus groups. Beth begins to outline the focus group process for him (see Chapter 6). She gives him an estimate of the time that would be involved to set up the study. She mentions serveral locations in the area where a focus group interview could be conducted. Marshall reveals that he is considering adding a new line of clothing, but he has reservations about how buyers will respond to it. Beth and Marshall spend the next 25 minutes discussing what could be unearthed from a focus group session and pricing for single or multiple interviews. Beth clarifies misconceptions about what a focus group will disclose. They examine possible scenarios for the interview, and Beth advises:

"Once a client discovers the intelligence harvested from these studies, they want to commission more of them for different types of buyers. I suggest that you begin by scheduling three sessions over a 12-week period."

Marshall is tempted by the idea but tells Beth that he thinks he'll begin with one focus group and wait to see what they learn. Beth reaches in her briefcase and retrieves a contract and her daily planner. As she begins to fill out the contract, Beth says:

"I believe we can schedule a session by the end of next month. I'll check with the facilities manager and our coordinator at the station."

Afterword

In the previous five chapters, the salespeople in these situations knew that some of the information they were presenting might not be news to the advertiser. They understood, however, that it was important for the advertiser to see that the salesperson had worked to earn the client's business. These salespeople also wanted to show their advertisers that they understood some of their needs. And, in doing so, they were establishing credibility with their clients.

Salespeople like the five highlighted here stand so far out of the line of sales-rep mediocrity that they are in another category. These salespeople didn't attempt to imply, or say, that their radio station was *The Answer*. If you'll notice, the majority of the time they weren't even discussing their station. The focus was on the advertisers' business and customers. They didn't suggest that they had all the answers, but showed that they would work with their advertisers as a true marketing partner—always trying to find new ways to help them increase their business. These are *true* niche marketing consultants.

15

Copywriting for Niches

The Importance of Questioning

Albert Einstein once said, "The important thing is not to stop questioning." If you write copy, you'll find it helpful to have that quote emblazoned over your desk in letters at least four inches high. Verbally question people who ask you to write an ad. You'll always get more information than what was written on a piece of paper. Just as an actor gives a better performance when the director has given her the motivation behind the character's actions, copywriters create more convincing ads when they understand what motivates a consumer.

- What is important to a consumer when considering the purchase of this product (or service)?
- What questions do customers ask when they're about to buy?
- What comments do sales associates hear from satisfied customers—and from dissatisfied consumers?
- How do you get inside a consumer's mind and connect with him or her on that "inner field"?
- How do you break through the nonstop barrage of information catapulted onto consumers every day of their lives?

Learn to Know Thy Consumer

Your First Contact

When a salesperson asks you to write an ad for a client, request a copy of the results of the Niche Marketing Analysis: Part I, the NMA: Part II, and any findings from customer surveys—either conducted by your radio station or an outside source. The most helpful information you'll glean from the NMA: I will be from the sections entitled Customer Profile, Positioning Analysis, and Advertising Goals. The entire NMA: Part II will help you gain more insights into what is important to this client's customers.

Create a file for each consistent advertiser and for each advertiser that a salesperson tells you he feels has upgrade potential. Each salesperson may have a targeted list of advertisers that will require more of an investment of your time. These are the clients—and their customers—you'll want to know intimately. Companies have personalities—usually like their owners. Consumers often buy from businesses that have personalities like their own. Therefore, when you understand the personality of an organization, you also know a lot about its customers.

Motives

To break through the profusion of information being thrown at shoppers every day, you need to get inside consumers' minds and draw them into your ads. In other words, you need to *become* your advertisers' customers. Learning the buying process from the customer's point of view may be the single most important action you take outside of actually writing the ad. You're already farther along the path to this knowledge than you think you are because *you're* already a customer. A good place to begin looking for a connection to your target consumer is inside *your* head. Maybe you haven't purchased the products or used some of the services for which you write ads; however, you have experienced the buying procedure for many products and services. Summon those experiences to help you begin to get in touch with the people you're writing to now.

The Essence. The essence of what motivates a person to buy is the goal for which you're striving each time you prepare to write an ad. In order to discover why people make purchases, it will help to understand basic and then the more advanced needs that humans possess. Abraham Maslow, a pioneering psychologist in classifying human needs, identified five categories of needs inherent in people (Figure 15.1). He begins with the most basic need for survival of the mortal body and goes on to the highest level or need for self-fulfillment and self-awareness.

The Missing Link. There is a direct correlation between people filling their needs and positioning a product or service in the mind of the customer to show how it will fill that need. Consumers will have a motivation to buy when they see a *direct link* between their needs being met and the retailer's products or services. Each category of needs corresponds with related products and services. To understand consumers, learn about their needs and the connecting types of related goods and services.

Physical Survival Needs

A healthy body, having enough food to survive, obtaining a shelter to protect the body, hospital emergency services, long-term caregiver services, repairs to the body from life-threatening diseases or injuries—all these would be filed under the most fundamental of human needs. Pharmaceu-

```
                                    | Self-
                                    | Fulfillment
                                    | Needs
                          | Status and
                          | Esteem
                          | Needs
              | Love and
              | Association
              | Needs
     | Safety and
     | Protection
     | Needs
| Physical
| Survival
| Needs
```

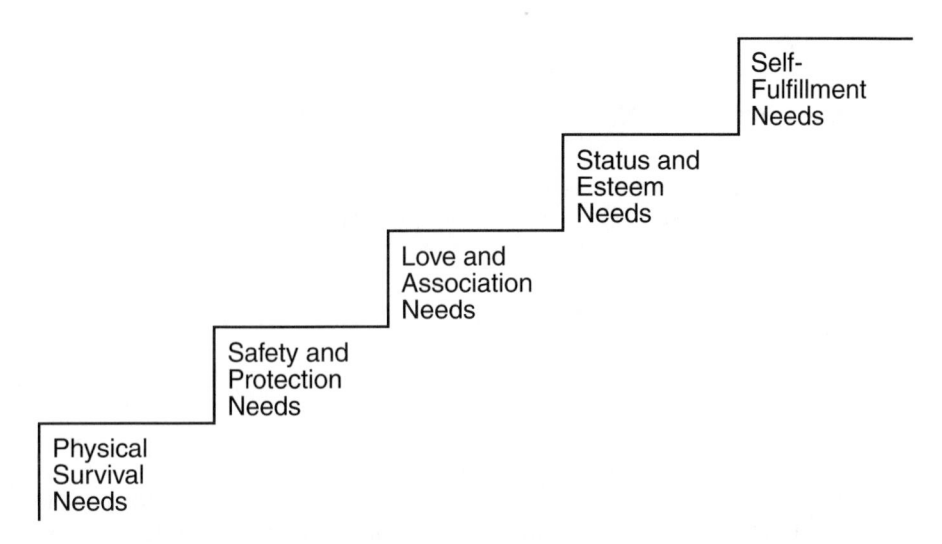

FIGURE 15.1 Abraham Maslow's Hierarchy of Needs

tical companies rarely advertise products that are necessary to avoid extinction. More often, drugs—either prescription or over-the-counter—are advertised for less severe physical ailments. This doesn't stop copywriters, however, from describing symptoms that *only this pill* will cure as if the consumer might collapse and their world may end if they don't take this special tablet.

As you know, physical needs for sustaining life also include the need for shelter and food. These needs would address the most fundamental shelters and primary need for sustenance—not luxury foods or condominiums, although some groups of consumers may consider luxury foods and condominiums *as* basics. When writing ads for products or services that fill elementary survival needs, the consumer won't make a connection if you're promoting your product other than as *necessary* for survival of the body. The majority of people in developed nations, however, find a way to fill these needs. Those who are still struggling for physical survival, food, and shelter usually feel powerless. Give them a sense of *not* being helpless, tell them how to solve their problem *now,* and your ads will have appeal for these consumers. (See Exhibit 15.1.)

Safety and Protection Needs

Once the most basic needs for survival are filled, people will switch their attention to filling their need for safety and protection. This level includes the need to feel safe from harm—for ourselves, our family, and our possessions. Consider the possessions of your target consumers. Do they live in an apartment? Are you discussing protection for the contents of that

Exhibit 15.1

Physical Survival Needs　　　　　　　　　　**Advertising Words** Some words that promote survival are: sustenance, life-giving, endurance, sustaining, strong, fortitude, stamina, rescue, fight, tolerate, persist, immunity, help, patience, lifesaving, and supportive. 　　　　*All day strong. All day long.* (Aleve pain-reliever)

apartment? Are they homeowners? Are you selling them homeowners' insurance? Do they have a family, a partner, or children? A security system for their home or automobile and smoke detectors would address this need. A functioning furnace or air conditioner would appear in this category in sections of the globe where severe weather extremes exist. Safety equipment for the home to avoid accidents or intrusion would be included in this grouping. Financial services to build and protect investments would fill a need for security. Preventive medicine and vitamins would fill a need for safety above the need for survival. (See Exhibit 15.2.)

Controlling One's Environment. Under the umbrella of safety and protection, you may add the need for order and the need to control your environment—which may be translated as having things happen in your life as you expect them to happen—to avoid uncertainty. The most direct connection to this need would be appeals to consumers promoting things that come in pairs or sets. Include stores with an expansive inventory of coordinated clothing—things that you can plan. Also appealing would be a well-designed service that is provided with regularity—something that furnishes consistency. Cleaning products or services for the home, car, and your body are in this group. People want to feel that they are in control—at least of their own lives.

Consumers have a need to control dirt and pests in their own home with cleaning aids and equipment such as vacuum cleaners, dusting aids, and pest-control products. Merchandise also includes water and fabric softeners, detergents, and personal hygiene products such as soap and shampoo. The need to control one's environment may apply to something as simple as predicting lawn care needs ahead of the season, buying clothing before the next season begins, or using contraceptive products. Within this category are sell service, satisfaction, and freedom from worry. (See Exhibit 15.3.)

Exhibit 15.2

Safety and Protection Needs   **Advertising Words**

To promote products and services in this category, use words such as—protect, safety, security, free-from-worry, defend, shield, safeguard, secure, cover, shelter, preserve, and maintain.

It's a good time to be silver. (Centrum vitamins)

First Alert. Every Minute Counts. (Smoke detector)

Dean Witter. We measure success one investor at a time. (Financial investments)

ADT. Security for life. (Security alarm)

Exhibit 15.3

Control (of One's Environment) Need   **Advertising Words**

Words that would appeal to consumers of products and services to control one's environment are: anticipate, foresee, envision, predict, control, peace, imagine, picture, expect, satisfy, pair, set, stable, balance, harmony, service, boss, worry-free, forecast, guarantee, collection, group, and match.

Raid kills bugs dead! (Insecticide)

Don't risk it. Wisk it. (Laundry detergent)

Eureka Bravo! The Boss. (Vacuum cleaner)

Bounty. The quilted quicker picker-upper. (Paper towels)

o.b. worry free. (Tampons)

Love and Association Needs

The third step on Maslow's staircase of needs is the need for love and a sense of belonging—whether that translates to a mate, friends, peers, co-workers, or strangers. Humans have a great need for association with others, that is, to feel love or, at least, acceptance. Products or services that make a person more attractive to others would belong in this category.

Drinks people enjoy while in the company of others—for example, coffee, soft drinks, and beer—are also included here. (See Exhibit 15.4.)

Exhibit 15.4

Love and Association Needs   **Advertising Words**

When writing ads to promote love and association, using words like admire, flatter, compliment, savor, enjoy, relish, value, admire, together, praise, love, honor, caress, cherish, awe, prize, care, treasure, applaud, celebrate, endorse, respect, and congratulate would connect with this need.

*When You **Care** Enough to Send the Very Best.* (Hallmark cards)

*Nothin' says **lovin'** like somethin' from the oven.* (Pillsbury cakes)

***Proud** to Be Your Bud.* (Budweiser beer)

*I won't dress before I **caress**.* (Caress body soap)

Nurturance. A grouping within the love and association need is the need to give comfort and support to others. This need may relate to people, animals, or plants. The need to nurture others may include one's children, mate, and relatives; and it may also embrace one's neighbors, employees, acquaintances, and even strangers. Animal lovers and horticulturists—often called *green-thumb* people—would be included in this group. Products and services associated with parenting, childcare, purchasing food, cooking, *family* laundry, lawn or garden products that promote healthy growth, domesticated as well as wild animals—in their natural environments—and pet supplies. Washing the family laundry isn't about the quality of cleaning products as much as it is about providing for the people who wear the clothes. Soliciting funds for nonprofit organizations is included on this list also. (See Exhibit 15.5.)

Sexual Individuality Need. Under the category of the need for love, humans have a need to establish their sexual individuality. As you know, they also want to feel attractive. A part of this need is to obtain and furnish sexual satisfaction. Women and men also want to whet their sexual appetites without being criticized for it. Therefore, products or services that promote these feelings would connect with people on their level of need for sexual individuality. Some of these are clothing and accessories

Exhibit 15.5

Nurturance Need	**Advertising Words**

When attempting to break into the consumer's mind at this level, employ words like care for, touch, nourish, feel, parent, adopt, sustain, cultivate, develop, save, reach, train, flourish, model, nurture, shape, and mold.

__Reach__ out and __touch__ someone. (Bell telephone)

__Save__ the Whales. (Greenpeace environmental group)

Makes a real good __feel good__ meal. (Hamburger Helper meal)

Recommended by doctors, pharmacists, and "__Dr. Mom.__" (Robitussin cough/cold medication)

that intensify sex appeal, such as perfumes and colognes, makeup (usually for females), hair-restoring products or surgery (usually for males), and intimate apparel for either gender. Books, tapes, and videos about sexual activity would also be a part of this group. Restaurants and bars that want to promote a marketing niche such as being a place where people meet, enjoy each other's company, and date would use words in their ads from the list in Exhibit 15.6.

Exhibit 15.6

Sexual Individuality Needs	**Advertising Words**

Some words that would help you capture the attention of listeners at this level are: enchanting, fascinating, delightful, delight in, inviting, charm, charisma, captivate, fantasy, arouse, excite, desire, tempting, appealing, attractive, singles, sexy, pleasure, encounter, luscious, lure, ultimate, hot, sensual, and ecstasy.

Ultima II (Lipstick)

Nothing comes between me and my Calvins. (Calvin Klein jeans)

The feel. (L'Oreal makeup)

Nothing beats a great pair of L'eggs. (Pantyhose)

Status and Esteem Needs

When writing to people who are searching for status and esteem needs to be met, remember that they will be asking themselves this question. "Will this (product or service) get me recognition or visibility?" Consumers trying to fill this need are looking for positive attention from other people to show their supremacy or excellence in a particular area. They've toiled long and hard to get what they've achieved and they're proud of it. Speak to them as experts because they're trying to impress others with their expertise. Products or services that show superiority, such as achieving educational degrees, awards, trophies, and special abilities, would appear in this grouping. One of the largest trophies some consumers buy is a home. Naturally, the status an automobile signifies appeals to this need. Associate jewelry with status to sell it to consumers attempting to impress others. Identify these prospects with famous people. Allude to being admired by others and you'll be appealing to people's need for status. (See Exhibit 15.7.)

Exhibit 15.7

Status and Esteem Needs   **Advertising Words**

Attract consumers at this level with words like expert, winner, overachiever, prestigious, professional, genius, master, specialist, champion, whiz, veteran, goal, ambitious, degree, power, status, capable, style, knowledge, experience, reputation, best, ruler, leader, owner, boss, captain, most valuable _____ , accomplished, convince, authority, supreme, victorious, win over, arrive, success, and achievement.

Phrases to attract this group include: Reward yourself. You've earned it. The Michael Jordan of _____ . The Garth Brooks of _____ . Now that you've arrived.

Be All You Can Be. (U.S. Army)

Master the Possibilities. (MasterCard credit card)

The Breakfast of Champions. (Wheaties cereal)

The Ultimate Driving Machine. (BMW automobile)

The Need for Independence. This need could be described as a sub-need after the need for status. Consumers who are trying to fill the need for independence want to be different. They want to be free from supervisors telling them what their next move should entail. Options and alternatives are very important to people at this level. Businesses that emphasize that their products or services will give buyers a sense of freedom, make them uncommon, and help them stand out from the crowd will be using appeals that attract customers trying to fill this need. Hair styling salons, foods that are not on the grocery list of mainstream society, customized products for the home, and customized automobiles are in this category. (See Exhibit 15.8.)

Exhibit 15.8

The Need for Independence **Advertising Words**

To sell your clients' products and services, use words in your ads like distinctive, special, uncommon(ly), extraordinary, individual, outstanding, unusual, only, rare, original, one, unconventional, detour, exceptional, stand out, and unique.

Jeep. There's only one. (Sport utility vehicle)

Saab. Find Your Own Road. (Automobile)

Recreation and Diversion Needs. At one time or another, humans feel the need for recreation and diversion. Consumers want something that stimulates their senses—whether that means performing some kind of physical activity or finding a change from their routine—and/or that entertains and relaxes them. This could be labeled a sub-need associated with the need for independence. Businesses that sell products that feature unusual scents, textures, or flavors will appeal to this group. Food with strong flavors, like Mexican foods, or foods with varied textures, like sushi, appear in this set. Theatre owners, recreational vehicle dealers, and sporting equipment store owners are some of the advertisers who may use this type of approach to attract customers. (See Exhibit 15.9.)

Self-Fulfillment Needs

The need for self-actualization has been called the ultimate need. When do consumers pursue the self-fulfillment need? People strive for these needs only after other requirements have been met because most people spend the majority of their lives trying to fulfill other, more basic needs. Understandably, it's later in their lives when they can pursue self-actualization. Consequently, products and services in this grouping appeal to a consumer who has achieved a degree of mental autonomy. Products that emphasize the ability to exercise one's skills and talents would attract a person at this level. Some would describe these consumers as having wisdom.

Exhibit 15.9

Recreation and Diversion Needs   **Advertising Words**

Words that catch the attention of consumers trying to fill recreation and diversion needs are: adventure, explore, challenge, stimulating, exciting, exhilarating, quest, relaxing, spark, play, kindle, stir the senses, fire, sport, ignite, change, amusing, alive, and feel.

Eat football. Sleep football. Drink Coca-Cola. (Soft drink)

Pure Chewing Satisfaction. (Wrigley gum)

Just Do It. (Nike shoes)

Filling self-fulfillment needs often means seeing the bigger picture. People pursuing self-actualization often think in terms of concepts rather than studying something step-by-step. These are *not* people seeking the spotlight for the sake of fame. They seek fulfillment for *their* gratification—or for the nonmaterial gifts they can give to others as a result of their continued search for knowledge. The more they learn, the less they feel they know.

These consumers often believe that exercising one's body is as important as exercising one's mind. Products related to advancement in one's profession or occupation, self-improvement programs, and physical fitness equipment fit into this category. Services such as cosmetic surgery, medically-supervised weight-reduction, charity marathons, and hair-replacement procedures appeal to these consumers. (See Exhibit 15.10.)

Use the Language

Whichever need you're addressing in your ad, show how your client's store, products, or promotion will fill that need. Now that you know some of the words of the consumer's language, use those words to help your listeners identify with your message.

Advertising Strategy

Analyze current ads and you'll hear in most of them only one uniform characteristic—they all say something favorable about the product. And, in most of them, you'll search—in vain—for real evidence of masterful generalship or *strategy* in influencing sales. Generalship in the battle for sales is as important as in directing armies. Tremendous forces of power may be brought onto the field; but unless they're directed with consum-

Exhibit 15.10

Self-Fulfillment Needs	Advertising Words

Recharge, relax, attainment, inner peace, solitude, pinnacle, ultimate, refuel, fulfill, reach, realize, evolve, how to, reflect, wisdom, intriguing, accomplish, achieve, and satisfy are advertising words consumers connect with when pursuing self-fulfillment and self-actualization needs.

Advertising phrases to insert in your ads include: the next level, not for everyone, make the most of, heal yourself, reach the apex, be the best you can be, the pinnacle of your career, the total picture, see the view from the mountain-top, teach yourself, a new realization, how to pick the right exercise class, give yourself the gift of _____ , the professional's guide, promise yourself, and self-exploration.

Jaguar. The Car of Your Dreams . . . At Last. (Automobile)

Take the Next Ten Years Off. la prairie (Skin care product)

mate strategy, they can wear themselves out against impregnable walls. If you write ads, you're in marketing. Generalship in marketing, however, means finding the motivation to purchase a particular product and directing your writing efforts there. The strategy that fits each client is different. What competitors are doing is not a *dependable* guide. They may be driving down an inappropriate roadway. You'll want to know what competitors are doing, however, so that you can help your client promote a niche their competitors are *not* filling.

Ask yourself, "What does my client want this advertising to accomplish? With few exceptions, the answer is "To create customers who will go into my stores (offices) and buy these products (services)." Advertising that accomplishes this will also remind current customers, *re*sell customers who have fallen away, and it may point out added uses to present buyers. It can also stir the enthusiasm of your client's salespeople.

First-Time Buyers

On the road to marketing generalship, you may be asked to write ads for the manufacturer of a new product. Your station's marketing consultants may be asked to create a marketing campaign for the new product. As you know, the chief aim of advertising is, without exception—to *create users*. The following list of techniques may give you new insights into the process and help you make contributions to the marketing team for such a project. Here are some of the methods frequently used to get consumers to use a product for the *first* time:

1. Offering samples
2. Offering booklets or further information
3. Offering premiums
4. Offering combination sales
5. Offering full-size packages free
6. Offering gifts to those who see demonstrations
7. Contests
8. Offering price concessions

Response

The preceding techniques can be used successfully with radio advertising. Use any of these methods well and you'll get some response. For any given product at any given time, however, one method always brings new customers at a lower cost than any other method. And, in each business, it may be different. Only by measuring the cost and results of each method and comparing it with the others, can the best strategy be found. You won't be working with your client's best interest in mind if you accept and use any one plan without knowing the comparative value of others. Generalship in copywriting requires learning these facts and discovering what actions to ask the listener to take. What benefits of shopping at that business are most influential in marketing that business on the air? Then, and only then, should the advertising be planned.

Strategy in advertising means finding sales opportunities that lie undiscovered when one uses routine advertising copy notes. It also unmasks the unprofitable ways that seem inviting on the surface before *true* buyer motivations are unearthed. An advertising strategy states—in the simplest terms—what the advertising should convey so that it is satisfactorily received and who should hear it. For example, a client's strategy may be:

1. Increase awareness of our 6 locations by increasing our advertising budget.
2. Offer demonstrations of our products to consumers who visit our sites.
3. Provide flex-pay options.
4. Provide professional consultants for our customers.

Face Facts

Some larger-unit products can't be sold profitably by using standard advertising methods. For years, without knowing this, many businesses beat their heads against an advertising barricade that was impenetrable. Others advertised, understanding the exorbitant cost and hoping that gross profits would be large enough to pay the bill. Now some of these businesses have found that advertising can be used profitably for products of this type if the *goal* is to sell people on seeing a *demonstration* instead of trying to induce people to go to stores just to buy the product. Offering a gift to bring qualified prospects in to see a demonstration in stores or showrooms can also be effective. This method has reduced the cost of selling sufficiently to pay the cost of the advertising and the free gifts and still leave a profit. It has also attracted countless prospects who could not have been enticed in any other way. Strategy is always more profitable than blind, brute effort.

Testing Appeals

In a midwestern city sits the president of a large food specialty company; he chuckles to himself whenever he hears the ads of his competitors. Both of his major competitors are using the same basic appeal in their ads—a different appeal than he is using. He knows that their appeal is comparatively inefficient. "My competitors must be using that appeal without knowing how it compares with others," he reasons. The midwestern president tested his company's appeal against four other appeals. He made tests by running five different ad campaigns on carefully selected radio stations over a period of time. He chose the most effective appeal because it produces sales at lower cost than any of the others.

Yes, the appeal his competitors use creates sales. Results show, however, that it takes about 38% *more* in advertising dollars to equal the results of the appeal he uses. When he hears his competitors' ads, he always smiles to himself. He knows that for every dollar he spends, his competitors are spending $1.38. Annually, he spends about $200,000 on radio advertising. He knows, however, that his competitors must spend about $276,000 to produce equivalent sales. The bottom line is that this businessperson has preempted his field. Competitors are now far behind.

Many businesses calculate the cost-per-inquiry of individual ads. These statistics, however, can't predetermine the results of new advertising unless the ads are written on a basic theme appeal from a niche marketing perspective. When an ad attempts to sell a product using clever words instead of a marketing plan, the successful ads are hard to duplicate because no one can tell which elements produced the results.

Making Your Job Easier

You may be wondering if you have time to delve this deeply for every ad you write. Indeed, you may not. Remember, when salespeople use niche marketing techniques to sell more advertising, they'll sell more advertising to fewer clients. You may be working with fewer advertisers. And, you'll have more knowledge about what drives people to buy a particular product or to do business with a specific company. This intelligence will actually make your job easier. Writing accurately niched copy is similar to closing a sale—the more time one spends qualifying before attempting to close a sale, the less time is necessary to close. The radio ad prep sheet is designed to help advertisers, salespeople, and copywriters.

Radio Ad Prep Sheet

The three-part worksheet in Exhibit 15.11 can easily be printed, padded, and given to advertisers who buy from your radio station regularly. Use this form to help you accomplish these six goals for your station:

1. The radio ad prep sheet (RAPS) can help an advertiser plan a more effective ad.
2. The advertiser will feel more involved in the creation of the ad. That feeling can promote a sense of control over what his advertising dollar is accomplishing. The client may also feel more a part of the *entire* process rather than someone who gives a sales rep information on a piece of paper. Some clients may have no further contact with the radio station until they receive a bill from it.
3. It's a constant reminder to an advertiser that the staff at your station is a *team* dedicated to helping him receive the best service possible. It also gives a glimpse into some of the enormous backstage efforts expended at the radio station to help air an effective ad.
4. The radio ad prep sheet will improve communications between your client, your sales rep, and you.
5. The RAPS makes an intangible transaction appear more tangible.
6. Your radio station will position itself as a professional entity with valuable knowledge that can benefit advertisers.

Copywriter Sleuthing

Successful radio salespeople invest a considerable amount of time learning about their clients and their clients' customers. Radio copywriters probably will never see a paragraph in their job description saying that they must learn about consumers. Don't allow the absence of an official request from management stop you from sleuthing. The most effective copywriters never end their quest for learning about the people to whom they write. Study surveys, opinion polls, and industry statistics. Read books about marketing, psychology, and sociology. Then, go among "them." Whenever you go into a restaurant—any restaurant—study the customers who are there. Ask yourself questions about these consumers. What are their age ranges? What income ranges do you think they fit into? Are most of the customers dining here families or singles? Then, look at the presentation of the food. Who is the restaurant targeting?

Exhibit 15.11

Radio Ad Prep Sheet

(Station logo)
(Station address)
(Phone number)

Today's Date: _____

This section to be completed by Advertiser:

Start Date for ad: _____
End Date for ad: _____

Business name: _____
Location(s)/Identifying landmark(s)/Exit #: _____

Phone number for ad approval: _____
Best time to call: _____
Contact: _____
Product(s)/Service to be advertised: _____

Biggest benefit: _____

Additional reasons to buy: _____

What problem is customer trying to solve with product/service? _____

Company theme or positioning statement: _____

Advertising goal for this ad: _____

Length of ad: ☐ 10-sec ☐ 30-sec ☐ 60-sec
Jingle to be used: ☐ Yes ☐ No
Specify cut #: _____ Length of lead: _____ Other: _____

Exhibit 15.11 *continued*

Today's Date: _____

This section to be completed by Marketing Consultant:

Approval needed by: _____
Start Date for ad: _____
End Date for ad: _____

I have delivered to copywriter: ☐ NMA: I   ☐ NMA: II   ☐ Customer surveys
☐ Focus group findings   ☐ Other: _____
Target customer demographics: _____

Lifestyle information: _____

Today's Date: _____

This section to be completed by Copywriter:

Approval needed by: _____
Start Date for ad: _____
End Date for ad: _____

What do listeners want when they buy this/these _____?
(products/services)

To which emotion can I appeal? _____

Which need am I trying to fill? _____

What is the consumer's point of view? _____

Will sound effects help this ad?   ☐ Yes   ☐ No   Specify: _____
Does client have a musical image?   ☐ No   ☐ Yes
  Do we have a copy in-house?   ☐ No   ☐ Yes
Specifics: _____

Allow your curiosity to take you into a store you've never visited. Choose a store in which you personally may have no interest. Look at the inventory. Who is buying it? What lifestyle are these products promoting? Which needs are this store's customers trying to fill? Talk to the people who serve consumers. Sales associates, executive assistants, bartenders, and receptionists interact daily with large numbers of consumers. They can reveal a wealth of knowledge about their customers. Mentally write an ad for each type of business.

If you have an opportunity to visit a business for which you write ads, quietly observe their customers. Do these consumers look like they're enjoying their visit? Are they leisurely strolling through the store or do they appear frustrated? Do you hear a number of people asking for assistance or is assistance offered before it is requested? Does the atmosphere feel warm and inviting? What questions do you hear buyers asking the employees? Is there anything positive about visiting this store that really seems to stand out? How does it differ from its competitors? Can you use that information effectively and positively in an ad?

In the beginning, use these techniques with your long-term clients. Invest more time and energy here. Use your efforts to help salespeople keep these advertisers and make them longer-term clients. Advertisers will be more successful because they'll be receiving more effective ads. Salespeople will be more successful and happier because they will be expending more of their energy developing productive campaigns instead of standing in that unending line of media reps selling spots. You'll feel more fulfilled and contented in your position because you'll be ensuring your value with the radio station and adding to its stability. Everyone wins.

Striving to Be the Best

Copywriters who strive to be the best need to expose themselves to the best advertising. Study ads produced by the best marketing firms and ad agencies in the world. Look, listen, and read. These ads are all around you right now. Once you begin to analyze consumer needs, you'll begin to see the thinking behind many of the ads you're exposed to every day. A new dimension of ideas will open to you, showing you how to help the companies with which your radio station works. Which need do you think the marketing people for Betty Crocker are addressing with their frosting ads? The copy says, "When you love someone, give them Betty Crocker frostings. Betty Crocker. Life is sweet." What is written *between* these lines? Do you think that buying Betty Crocker frostings *really* shows love? Perhaps yes. Perhaps no. Ries and Trout (1981), authors of *Positioning: The Battle for Your Mind*, explain it something like this: It's not what you do to the product but what you do to the mind of the prospect. Some may say, in marketing, perspective is everything.

Becoming Aware

Perhaps because of the volume of writing a copywriter is asked to produce, many times the focus is on telling—or yelling—that a certain product can be obtained for *90 percent off for the next two hours at only this location*. Indeed, many retailers have reacted to increased competition by lowering prices on their inventory. That leaves radio stations in the position of being a fast-acting clearance-sale advertising medium. Change starts with you. "But what can I do?" you ask. "I'm only a copywriter. Nobody answers to me. I don't set policy at my station." Copywriters may not be on the top rung of the management ladder. You are in control, however, of your level of expertise. The first step is awareness.

Build a Reference Library. When you're looking through magazines, notice the slogans in ads. Tear ads out of specialty magazines and file them, noting in which magazine an ad appeared. Describe the target reader for that magazine. How does that ad fit into the lifestyle of the targeted reader? Build a collection of files based on Maslow's needs hierarchy.

Watch television. Yes, watch television. At least focus on the ads you'll see—especially the national ads. Many are from top-notch ad agencies. Look at the format of the ad. Watch for the theme. Where is it placed in the commercial? What are they trying to say? What is said between the lines? Who is the targeted viewer? How does the product or service fit into their lifestyle? Is the target buyer the obvious consumer or an influencer—a person who influences someone else to make a particular buying decision? Some influencers are kids who tell mom which cereals to buy. Other purchase influencers are males who tell their female companions which lingerie to buy. Keep a pad of paper and pen near you so you can record the slogans and copy written by leading ad agencies. Put these in your growing files.

Listen to other radio stations. Listen to national ads on your radio station. Take notes. Add them to your files. Soon, you'll have files bulging with phrases, themes, slogans, and advertising words. Your library of files will help your creative juices flood your brain with ideas during dry spells. You know, when you're beginning to knock on the top of your head and say to yourself, "Hel-lo-o! Is anybody home?!"

Cookie-cutter Copywriters

Radio station copywriters are usually asked to write ads that invite consumers into a retail outlet to buy merchandise or into an office to purchase services. Selling a particular product is usually left to the manufacturer on a national or regional level. If a hardware store owner wants to promote Stihl chainsaws, she usually wants an ad that simply tells listeners that the saws are available at her location. If competition is strong, she may add a sale price to entice a consumer to buy from her rather than anywhere else. She often depends on the manufacturer to create interest and desire with national advertising campaigns. Her vendors frequently tell her about the national advertising support she'll receive when she carries their brands. It makes sense, she reasons, to spend her 30 or 60 seconds telling customers about her great *l o o o w* prices. Write this kind of ad and she'll probably be happy with you. She'll also be happy with your competitors the next time she wants to run a similar ad and she feels it's *their* turn to get her advertising buy. "That's just the way it is in my market. There are a lot of stations," you explain. In the niche marketing future, however, only marketing-savvy copywriters will enjoy job security. They'll also enjoy their careers more.

Stepping Out of Line

Never stop sleuthing. Study people. Study service. Study the layout of merchandise in stores. What is the store manager trying to get people to notice the most? Where does the floor plan lead customers? What does the inventory tell you about the lifestyle of the targeted consumers? Go to a mall and visit many different kinds of stores. What do they have in common? Which stores make you feel the most comfortable? Why?

Understanding the Inner Circle

A person's clothing is a billboard for his or her opinions, viewpoints, and lifestyle. Look around you. Can you guess the type of consumers who would enjoy seeing a particular movie that's playing at one of your advertiser's theaters? Which words or phrases would these prospective moviegoers use in conversations with their friends? How would they review the movie for their friends? What phrases or words would they use to recommend it? Regardless of the format at your radio station, many listeners have something in common with you. Think about those commonalities. Use that knowledge to mentally reach out and talk with them as you would to people in your inner circle of friends.

Business as Usual

A copywriter at a radio station on the West Coast needed to write an ad for a carpet store. The manager of the store had an excess of carpet remnants. He decided to have a one-week sale to clear out as many remnants as possible. Chelsea, the copywriter, could have written an ad telling listeners that there was an incredible sale of remnants at Seal's Carpet Store. The manager would have felt that Chelsea and the radio station had satisfactorily completed the task.

Applying Your Insights. Chelsea, however, had spent a lot of time studying people and their buying habits. The copywriter thought about the people who would be interested in carpet remnants. Chelsea wondered how *she* would tell a friend about the sale. She pondered the target customers—they probably are bargain hunters, she reasoned. Antique collectors are often bargain hunters who fantasize about making the great discovery, she thought. They picture themselves strolling through the back of a dusty antique shop where they discover *a find*. Then, they imagine themselves as they nonchalantly amble to the cash register to buy the item before the owner realizes the object is really a rare collectible. Once out of site of the antique shop, they jump for joy and congratulate themselves for the astute collector they've become. Chelsea took the example of the antique collector's fantasy and applied it to the carpet-remnant buyer. In her ad, two female friends are talking on the telephone (Exhibit 15.12).

Exhibit 15.12

> SFX: (*telephone rings*)
>
> VOICE 1: Hello.
>
> VOICE 2: (*whisper*) Cindy! They've made a huge mistake at Seal's Carpet Store!
>
> VOICE 1: Linda, is that you? *Who's* made a big mistake?
>
> VOICE 2: (*normal but excited voice*) Seals Carpet Store is having a carpet remnant sale . . . but they've got the **wrong** tags on the remnant pieces! I saw it with my own eyes! The sign says **remnants** . . . but some of the carpets are *huge!* I could hardly believe it! . . . I looked around to see if anybody had noticed that they'd made a mistake! I saw a 12 by 12 carpet for $99! I didn't waste any time! I got one of the 12 by 12 foot carpets for Tommy's room . . . and I found the softest shade of blue for that long hallway upstairs. Cindy . . . drop *everything* you're doing and *get* over to Seals . . . before they find out that they've put those low prices on those huge carpet remnants! You know . . . Seal's Carpet Store . . . over on Rivers Ave. . . . at the Elm Street exit.
>
> SFX: (*dial tone*)
>
> VOICE 2: Cindy! . . . Cindy!

Listeners Respond. The ad drew bargain hunters within the first hour that the ads ran. People also took the store manager aside and confided that they had heard that certain carpet pieces were incorrectly marked. The ad was an overwhelming success for the client. Not only did he move unwanted inventory, he sold regular carpets as well.

Bending the Rules. The above ad broke several rules of ad writing. First, Chelsea used pronouns several times in the ad instead of using the client's business name. Second, she didn't say the client's name at least four times in a 60-second ad. Third, Chelsea did not say the client's location more than once. Experienced marketers realize that a level of name awareness will be developed in a market over time. If the advertiser's name recognition is high, shortcuts *can* be taken and rules for writing *can* be bent. Sometimes, a business will be given a nickname—perhaps it could be called a niche name—by its customers. With high name awareness, marketers can identify their client with this nickname instead of the advertiser's full business name and still be successful. Knowing when and how to bend the rules comes only after much experience with success *and* failure. Proceed with caution.

Becoming a Good Self-Critic

To achieve generalship in marketing, develop the habit of becoming a good self-critic. After you've written an ad, ask for input from the sales rep, the program director, and the general manager. Survival training teaches that two heads are better than one and three heads are better than two, and so on and so on. When someone shares her perspective, don't close your mind to it. First, say thanks for it. Second, try to absorb the new information from the other person's viewpoint. In the end, you may choose to use the new suggestions or you may reject them. Even though the process of opening yourself to someone else's point of view may be painful, it will make you a better copywriter.

Who Has THE Answer?

We, the authors, want to assure you that *no one* has *the answer* to what motivates consumers. To be the most effective, however, your ads must convey that you understand the listener and that you speak the same language. Your greatest achievement will occur when a consumer hears your ad and silently says, "Somebody *understands* me!"

> **A Note to Management**
> Don't ask your copywriter to be a copywriter/receptionist. Insightful, niched copy is too important to the success of your advertisers and your radio station. Take every step possible to foster creativity within your copywriters. Educate them. Encourage them. Include them in strategy sessions for clients. Smile. Your competitors don't even think they *need* a copywriter.

PART FIVE
Clients as Partners and Looking Ahead

16

Partnering with Ad Agencies

Advertising Agents and Agencies

Circa 1888, printing in mass quantities became easier and cheaper due to technical advances in the production of paper and more efficient presses. As a result, newspapers and magazines flourished. Businesspeople fell in love with this inexpensive way to reach mass audiences. Potential advertisers, however, had few reliable ways to measure actual circulation. There were no central sources that told them how and where to find the publications that were available. This scenario inspired people like George P. Rowell, who developed the *Advertising Agency Circular* in New York City—a monthly magazine that listed available publications and their *reported* circulation. Thus, George became an early ad agent. George made money by selling subscriptions to businesses interested in advertising, and his agency bought and sold large blocks of newspaper and magazine space. Rowell's idea helped him make a fortune, which naturally attracted hundreds of imitators.

Competition Breeds Innovation

Circa 1915, some of the early ad agencies, like Lord and Thomas in Chicago and McKim in Toronto, were busy buying large blocks of newspaper space. They would then broker this space at a 15%—or better—mark-up to their clients. Agency managers discovered that clients were asking for and *would pay* for help with their copywriting, headlines, and print layouts. Early ad agency people like Claude Hopkins and Albert Lasker developed new techniques for writing powerful headlines and body copy. The agency business began to grow by expanding their menu of services. The love affair with newspapers and print still lingers in some old-line ad agencies. This may explain, in part, why some North American newspapers still enjoy more ad revenue in their respective markets than radio, cable, and television combined. The advent of niche marketing can help change this mass-market advertising tradition.

The Imputes for Mass Media

The market demands for mass media were fueled by the smokestack economies of the early twentieth century. Ad agencies were quick to adapt to this mass-market revolution. Advertisers who mass-produced goods and services felt quite comfortable with large newspaper circulation figures. So, large newspapers, such as the *New York Post* and the London *Times*, flourished in the 1930s. Mass-market radio stations like KDKA in Pittsburgh and WMAQ in Chicago, which covers 38 states at night, prospered in the 1940s. Network television stations like KABC in Los Angeles enjoyed record ratings and revenues in the 1950s. Ad agencies, eager for mass-audience numbers, helped foster what seemed like never-ending television growth in the 1960s and 1970s. The proliferation of media choices and the explosive growth of cable television caused some ad agencies to give radio sales and marketing presentations in the 1980s and early 1990s a low priority. Niche marketing, however, is changing the way ad agencies do business.

Mass-Market Media Rationale

For the past four decades, most ad agencies have tried to generate mass numbers of ad impressions. They also have tested ads for recall scores. Newspapers and magazines helped pioneer the idea of advertising recall with the Starch Reports (see Figure 16.1). Ad agencies used focus groups to test radio and television recall. Product recall was high on the list of desirable client objectives. Things, however, are changing. Today, ad agency executives realize that recall does not necessarily mean sales.

IBM provides a classic example of why niche marketing is replacing mass marketing. IBM has approximately a 98% recall factor in North America, Japan, and Europe (Figure 16.2). IBM sales, however, continue to decline. They're losing share-of-customer as well as share-of-market. They are being niched-to-death by countless competitors who offer compatible products with advantages other than just price. These competitors join the market quickly. They need market research today and advertising tomorrow. New companies, like ad agencies—especially large agencies—rely on ad-recall studies. These studies help justify mass-market ad campaigns and media placement decisions. Smaller agencies without cumbersome bureaucracies may be better prospects for niche marketing presentations.

Ad Agency Giants

Only a handful of ad agencies in existence today can be called full-service—that is, agencies that will handle the many facets of a client's research, advertising, public relations, media, and production needs. Agencies like J. Walter Thompson and McCann-Erickson, Inc., qualify as full-service shops. Full-service agencies offer a large menu of services. You'll see these

MAJOR NEW REPORT ON STARCH READERSHIP STUDIES OF 2,000 NEWSPAPER ADS

From one of the largest compilations of readership data ever assembled, guidance for advertisers on making more effective use of their media dollars.

Starch INRA Hooper, nationally known research firm and specialist in print ad media documented what happens when an ad appears in representative newspapers all over America. The results of these studies and how they can help advertisers are summarized in this new report. It includes readership of more than 30 advertising categories.

FIGURE 16.1 *Starch Report*—A print advertising recall study. From the Radio Advertising Bureau, 1988.

giants engaged in public relations; performing research functions; conducting media planning and buying; creating original layout concepts; buying finished artwork; supervising print and electronic media production; creating copy concepts; and writing finished texts for billboards, newspapers, direct mail, catalogs, magazines, radio, and television ads.

These huge agencies produce commercials around the world. They maintain service offices in most major cities across planet earth. Are they complex? The important thing to remember is that agency executives put

FIGURE 16.2 Consumers report 98% recall of IBM logo. From *Broadcasting and Cable*, July 1994.

their pants on just like everyone else—one leg at a time. They possess no magic. And, most will listen to intelligent niche marketing radio presentations, which are loaded with local marketing intelligence, that help their clients sell goods and services. These mega-agencies, as well as local retailers, have an unquenchable thirst for information about consumers.

Hot Shops

At the other end of the ad agency spectrum, you'll discover "hot creative shops" that specialize in creating copy, art layouts, and television storyboards for their clients. All other services are farmed out or performed by their clients. You'll probably encounter many agencies that image themselves as "hot creative shops." This does not always bode well for radio because the creative people at agencies can make flashier presentations using print and television. Niche marketing, with its narrow consumer focus, will help radio compete with the "hot shop" tendency to create mostly flashy mass-appeal television and print ads because mass-appeal ads will continue to diminish in their effectiveness. Radio, on the other hand, will continue to gain strength by focusing on niches that produce solid results.

The Agency Red Book

Many medium-sized ad agencies specialize in a certain type of account, such as banking, financial services, or package goods. Expect to encounter agencies that specialize in magazines and newspapers and place most of their advertising in print media. Some agencies specialize in the creation and placement of television. Very few agencies are radio specialists. The fastest way to find out about the types of accounts and the media placement patterns of most reputable ad agencies is to consult the pages of "The Ad Agency Red Book."[1] Many local libraries carry a copy of it. At a minimum, you'll get some idea about the kind of accounts each listed agency serves.

Agencies You Won't Find in the Red Book. The bane of many radio stations is the one-person ad agency. These countless enterprises are everywhere and exist in the largest and smallest markets. They rarely offer copy, production, or public relations services. Like the original agencies of the 1920s, their principal function is to negotiate for media time and space. These small shops often play one local radio station against another for "the best copy," or "the best campaign or contest idea." Then, they present the ideas to their client as their own. Many small shops commit a cardinal advertising and marketing sin—they allow each station to run their own copy and campaign. What a wonderful way to confuse the consumer in an age in which advertising clutter is rampant. The best tactic against this lose–lose strategy is to develop outside sources for copy, artwork, and production. Then, offer these professional resources—at a 17.65% mark-up—to clients who retain one-person agencies. You'll develop stronger relationships with your clients and serve them better than anyone else in your market.

Media Buyers a.k.a. Negotiators

Smaller specialty shops—ad agencies—that offer little more than creative services, such as art, copy, or layout, sometimes hire media buying services. These buying services employ people who are trained to be skilled negotiators. The following are some of the techniques buying service people are taught to use when they are dealing with radio salespeople:

   1. Radio time is perishable. Consequently, station representatives will offer rate and merchandising concessions so that their station will be included in a buy.

   2. There is high turnover in radio stations' sales departments. Therefore, many radio reps have few negotiating skills and they're constantly

---

[1] *The Standard Directory of Advertising Agencies*, known in the industry as "The Ad Agency Red Book," is a multi-volume directory published yearly by National Register Publishing, a Reed Reference Publishing Company, located in New Providence, NJ. A yearly subscription costs approximately $575. You may, however, find a copy in your local library; or, you can access the information you need on CD-ROM by calling 1-800-521-8110.

being pressured by management to get sales. In this stressful atmosphere, they become emotionally involved, which leads to more rate concessions and bonuses—like free remote broadcasts.

3. Many radio reps have little or no knowledge about the field of ratings. The best they can do is rely on computer print-outs and programs to analyze audience data. Many radio reps don't know how to calculate elementary ratings information such as cost per point (CPP) and cost per thousand (CPM). Some agency buyers have been known to slightly alter the demographics requested to confuse the rep during a negotiation in order to extract more concessions.

4. The confidence level of an *unskilled* radio salesperson is often very low. They are not very knowledgeable about their product. In addition, they usually have not studied psychology, how to read body language, or the basics of dressing for success. Skilled media negotiators use this lack of knowledge to gain concessions.

Using the presentation skills outlined in Chapters 10 through 14 can help you overcome these popular media buyer negotiating tactics. At the buyer level, the following section about the Media Buyer Needs Analysis will also help.

Media Buyer Needs Analysis

Here's a way to overcome some of the techniques ad agency buyers and media buying services use when dealing with radio salespeople. The Media Buyer Needs Analysis can be used for the times when they call your station and ask about avails and rates. When the buyer calls, you don't have much input about the decision making for the buy. At that point, the media planners have pretty much decided who will be called and, based on negotiating rates and rate *concessions*, they have a pretty clear picture of which stations will get the buy. That is *not* where you want to make an entrance with this advertising campaign. This is called "being at the wrong end of the food chain!" Keep a form like the one in Exhibit 16.1 near your phone, however, so you can get information to make a better last-minute or of-the-moment presentation. Use the form to help you get the information you need to make a short phone or fax presentation. Remember, *details* are important to buyers, planners, and account executives. It's their responsibility to justify their media buying decisions to their client. The more data you have, the more you'll be able to help media buyers justify those decisions. Plus, in the long run, you'll enjoy more business.

Getting Caught in Traffic. The Media Buyer Needs Analysis is designed for the eleventh hour. It's most helpful when you receive a buyer's phone or fax request for advertising availabilities. The information you

Exhibit 16.1

Media Buyer Needs Analysis

Today's date: _____

Agency name: _____
Buyer's name: _____
Agency address: _____
Phone number: _____ Fax number: _____
Advertiser: _____
Advertiser address/locations: _____

1. When do you plan to place an order? _____
2. What are the flight dates? _____
3. Which dayparts are you considering? _____
   ☐ Monday–Friday only   ☐ Weekends   ☐ Monday–Saturday
   ☐ Monday–Sunday   ☐ Morning drive—6 AM–10 AM
   ☐ Mid-day—10 AM–3 PM   ☐ Afternoon drive—3 PM–7 PM
   ☐ Evenings—7 PM–Midnight   ☐ Overnights—Midnight–6 AM
4. What's your primary demo? _____ Secondary demo? _____
5. Any ethnic weighting?   ☐ Yes   ☐ No   If yes, what are they? _____
6. What is your radio budget? _____
7. How many stations are you planning to buy? _____
8. Which stations? (fill in call-letters) _____
9. Which syndicated ratings (if any) are you using? _____
10. Do you have a target cost per point (CPP)? _____
11. Do you have a target cost per thousand (CPM)? _____
12. What other criteria will you be using to determine your investment? (for example, format? client preference? ad rate?) _____
    _____
13. How many gross rating points (GRPs) are you buying? _____
14. How many GRPs are you buying each week? _____
15. May we create a special package for you? _____
16. If we can meet your criteria, is there any reason why (fill in your call letters) would not be considered? _____
17. Media planner's name: _____
18. Account executive's name: _____
19. Before you finalize this buy, if there's any problem, would you give me the opportunity to discuss the problem with you?   ☐ Yes   ☐ No
20. What questions should I have asked that I did not ask? _____
    _____
21. When should I get back to you? _____

glean will help you with last-minute negotiations. Many radio sales representatives are reactive. They wait to be called by a buyer. They make symbolic agency calls once or twice a year on agency buyers who have little decision-making powers. They fail to realize that agency buyers—especially in large agencies—may be dealing with several hundred markets at one time. Given an average of ten viable radio stations per market, that could mean as many as 2,000 presentations per buy. The authors would like to help you avoid this traffic jam. When you deal exclusively with media buyers, you receive only minutes to make your presentation, at best. The long-term alternative is to consider a new avenue—working with the planner or account executive.

Avoiding the Traffic Jam. In the go–go years of the 1980s, when many stations were owned for 18 months or less, short-term selling strategies became very popular. People purchased stations for the short haul. Leveraged buyouts and easier credit markets made media property turnover possible. Calling on only media buyers and fighting for every penny in a last-minute buy made sense in the 1980s—*Get the billing up at any cost* was the theme. Today, with the advent of multiple station ownership in fragmenting markets, stations have to be held longer. The financial markets are forcing owners to build multi-station complexes in many medium-to-large markets. The exit strategies are based on holding stations for 5 to 7 years instead of 1 or 2 years. In a report developed by Hartsone & Dickstein, Inc., an investment banking firm specializing in broadcast properties, you can learn about *seasoned cash flow*, which means a track record of profits and cash flow that lasts for 4 or 5 years. Media selling strategies need to catch up with these harsh new financial realities. You need to build more long-term relations with your advertisers and agencies.

The Media Planner Needs Analysis

Building a relationship with a media planner can take several years and a good deal of research. Exhibit 16.2 shows an advanced needs analysis form designed to be used at the agency planner and account executive levels. The media planner at most agencies is the person who focuses on understanding the client's consumers and then sharing that information with the rest of the agency team working on that account. Your objective is to learn as much as you can about *their* clients and *their* clients' customers as possible. Then, you can develop an effective advertising strategy for using your station to make the client successful. Your job is performed *before* the actual advertising media plan has been developed. This means you may be selling 12 or more months before the actual advertising buy is placed. What follows will help stations receive a much

Exhibit 16.2

Media Planner Needs Analysis

Agency name: _____ Date: _____
Agency address: _____

E-mail address: _____ Phone: _____ Fax: _____
Advertiser name: _____
Advertiser address/locations: _____

Planner's name: _____
(for this account)

CONSUMER PROFILE

Heavy user demographics: ☐ 12–17 ☐ 18–24 ☐ 25–34 ☐ 35–44
☐ 45–54 ☐ 55–64 ☐ 65–74 ☐ 75–84 ☐ 85+

Buying decisions: ☐ Females ☐ Males ☐ Couples
Lifestyle(s) of your client's heavy users: _____
Heavy user income: _____ Heavy user education: _____
Other attributes of your client's heavy users: _____
How far do (your client's) customers travel to get to the store? _____
Are your client's products (services) impulse buys? _____
Why do customers shop there? _____
Average frequency of consumption:
   ____ Times daily ____ Times weekly ____ Times monthly
   ☐ Other: _____
Account's consumption objectives: _____
Are there specific market conditions the agency looks for when promoting your client's product (service)? _____

CONSUMER ATTITUDES

When people think of your client's company name, what image do you want to come to mind? _____

What is the number one misconception customers have about your client's company? _____

What are the most important benefits of shopping at your client's business?
_____

Exhibit 16.2 *continued*

MARKET RESEARCH

Has your agency conducted: ☐ Focus groups ☐ In-person interviews
☐ In-store surveys ☐ Telephone interviews
☐ Other: _____

Principal Competition

Who are your client's largest competitors? _____
  Why? _____
Have your client's competitors changed much within the last few years? _____
  How? _____
What makes your client different from competitors? _____

ADVERTISING PROFILE

Advertising objective: _____
Which media do you buy for this client? _____
(go to appropriate section)

Newspaper: _____
(list papers here)
Why is newspaper advertising effective for your client's business? _____

Are there disadvantages? _____
Yellow Pages: _____
(list phone books here)
Why is yellow pages advertising effective for your client's business? _____

Are there disadvantages? _____
Radio: _____
(list radio stations here)
Why is radio advertising effective for your client's business? _____

Are there disadvantages? _____
Cable/Over-the-Air Television: _____
(list television stations here)
Why is television advertising effective for your client's business? _____

Are there disadvantages? _____
Direct Mail: _____
(list when direct mail sent out)

Why is direct-mail advertising effective for your client's business?_____

_____

Are there disadvantages?_____

Do you combine your direct-mail efforts with other media?_____

How do you generate your mailing list?_____

How often do you update your list?_____

Do you purchase your mailing lists locally?_____

Other: _____
(e.g., billboards, inserts, magazines)

Why is _____ advertising effective for your business?_____

_____

Are there disadvantages?_____

Advertising Objective

What is your advertising objective for this client?_____

When do you make advertising decisions?_____

(Ex:) ☐ When the need arises ☐ Weekly ☐ Monthly ☐ Quarterly
 ☐ Yearly ☐ Other: _____

Presentation Appointment

What would you like from media reps that you're not getting now?_____

_____

Are there any other questions I should have asked?_____

_____

What do you need from me, as a media rep?_____

_____

I need approximately seven days to conduct my in-house research and prepare a presentation. Is this time next week good for you?_____

Date: _____ Time: _____

larger piece of the advertising budget than the traditional 7% to 10% most radio markets receive today. The business will last longer and larger profits will be the end result.

Success with Planners

The information gleaned from a Media Planner Needs Analysis will help separate your sales force from the legions of media peddlers who call on ad agencies. The Good Ole Boy (GOB) and Good Ole Girl (GOG) techniques of schmoozing the media buyer will not be tolerated at the media planner and/or account executive level. The new breed of radio niche marketing consultants will need to understand clients' businesses and their advertising and marketing strategies. The shallow, *I'm-number-one* tactics used with media buyers have no place here. Also, be aware of how pressed planners and account executives are for time as ad agencies downsize their staff to save money. Remember, the entire inventory of an ad agency walks out every night and returns, hopefully, the morning of the next workday.

The Word According to Courtney

In a January 28, 1994 *Radio Ink* article, Judy Courtney, senior vice president/media director for Foote Cone Belding Advertising in San Francisco, urged radio salespeople to call on ad agency planners. This strategy has been reinforced by the experience of the authors. We've never encountered a media planner who was unwilling to see a radio representative. They do, however, insist that the material presented not be audience numbers the agency already has in their computers. The strategy is to learn the agency's advertising needs and objectives for that client, then to present current market and station information that will help the agency meet the needs of the client. Don't hesitate to present niche marketing ideas combined with station events, promotions, or the offer to conduct customized customer surveys for the agency—unless you know that the agency conducts its own local, customized research (unlikely in most cases). Build the cost of the survey into the advertising package. *Never* offer to conduct the research for free or as a bonus. You will be saying that your services have no value. And you will be destroying your superb credibility.

Offering to work with the media planner to uncover consumer information will make you and your station outstanding in the planner's mind. Remember, everyone on the agency team wants to be a part of a winning campaign. You can help them achieve their goals by offering services—like customized, local research—that will aid them in learning about their clients' customers. Being able to spot a need that is not being filled and then responding before anyone else, can help your partner, the ad agency, keep *their* client. And, it can create the competitive edge for that client, which translates to continuing business for *both* of you. Your reputation as a knowledgeable niche marketing consultant will soar. Business will follow.

When to Go to the Client

Good media directors will even encourage station representatives to call on their clients—after the media buyer says, "No." Or, when the agency representative doesn't know the answers to all your questions on a needs analysis form. Ask permission to call the client and set up an interview in order to complete the research you need to make your presentation to the agency. The rule for calling directly on advertisers is to keep the ad agency informed about the details of your visit to the ad agency's client. It's a professional courtesy. It's essential for long-term client–agency–station relationships. As you enter the twenty-first century, your objectives *may* be diametrically opposed to the objectives of the ad agency. Many agencies are still trying to build share-of-market. Your niche and one-to-one marketing objectives for advertisers will evolve into building share-of-customer. Niche marketers will concentrate on how to increase consumption by their heavy users. Your job is to educate, educate, educate.

Share-of-Customer

The marketing focus of those companies that are moving to niche and one-to-one marketing is on share-of-customer—instead of share-of-market. Today, Pepsi-Cola sells more units of soft drink than Coca-Cola. Coke, however, still makes more money than Pepsi because Coke consumers drink more gallons per customer than Pepsi consumers. The Media Planner Needs Analysis is designed to ask questions about heavy users and what motivates them to buy. If you can show reasons why your market (audience) consumes great quantities of a client's product, you'll be in a better position to receive the radio buy.

Consumption habits of your audience is the kind of information not found in a ratings book. Services like Simmons give broad consumption figures for radio station formats. We're suggesting that you invest the kind of money and effort that people like Skip Finley of WKYS-FM in Washington, D.C. invests. By tracking bar codes at checkout counters, he can tell which products are heavily consumed by his audience—and in which stores—in his coverage area. Imagine the business he generates from national and regional brands that other radio stations do not receive. Also, there are hidden factors that influence agency media buying decisions.

Challenges, Creativity, and Technology

Production Challenges

Ad agencies usually mark-up all audio, video, and print production 15% to 20%. The average cost of producing a 30-second television commercial is approximately $100,000. The average cost of creating a 60-second radio ad is about $1,000 to $2,000, not including jingles and union talent when produced in a major market studio. Which medium would you recommend if you were in the 15% to 20% mark-up business? Using niche marketing techniques will help change the traditional television production advantage. It can also take weeks to write an ad, gain client approval, and produce a television ad. Good radio copy and production from ad agencies like Young and Rubicam can be written, approved, and produced in less than two weeks.

Creative Challenges From Confucius

Imagine being an ambitious copywriter or artist at a major ad agency. You've finally landed a dream job. You want to move ahead. You need samples of your work. Print and television give you an accepted showcase. Playing an audiotape isn't very sexy. We live in a world that harbors many misconceptions about words. Jack Trout, president of Trout and Ries Marketing Strategists in Connecticut, talks about one of the great myths radio faces: "A picture is worth a thousand words." The correct translation of the Confucian saying is: "One picture is worth a thousand pieces of gold." Take the same translation liberties, but be more incisive. Change the saying to "One word is worth a thousand pictures." Radio people have had to live with the inaccurate Confucian translation for decades.

Radio—Making It Glamorous

The creative talents at many agencies shy away from radio because they don't see it as a glamorous medium for their sample case. This is one more hidden reality radio faces when trying to gain share-of-mind and advertising budgets from ad agencies. The Radio Mercury Awards presented by the Radio Advertising Bureau (RAB) are helping to slowly change the perception that radio isn't glamorous—or profitable. Imagine being able to say that you just won $250,000 for creating a radio commercial. Your national sales manager (NSM) might want to take time to find out who the writers are on a specific agency account's creative team. Then, your NSM may take a few minutes to contact the writer or writers by phone and conduct a short, informal interview asking about the creative objectives for a particular account.

Radio's Smart Cards

There's another hidden objection to planning radio campaigns. Ad agencies we've surveyed are enthralled with interactive television. They see this technology as the answer to fragmented media. They expect to target consumers with a combination of televisions, computers, and direct mail. The technology for radio broadcast data systems will be truly interactive—even in your automobile; the hardware should be available by 1998. Mass distribution will probably be in place by 2001.

The system is simple. If you hear an ad you like, you can push a button on your radio and a coupon will be printed or recorded on a Smart Card. Take the coupon or card to a retailer and redeem it. You'll also encounter radios with L.E.D. displays that show the name and address of an advertiser. You'll be able to see the name of the song being played. You can then record the name of the song on a Smart Card. Take the Smart Card to a retailer with a scanner, and you'll be able to buy the CD. This kind of interactive technology needs exposure at the ad agency planner and account executive levels. Radio is ready to interact with the niched demands of the future.

Radio has been a one-to-one interactive medium with 800 numbers for years. A live person at the other end of an 800 number, however, may be too pedestrian for many agency executives who do not understand the complexities of interactive technologies, so far. Take heart. You've just examined some of the challenges radio stations face with some agencies. There are still many ad agencies that see radio as a very important medium. They are committed to working with stations for the benefit of their clients.

Agencies of the Future

Agencies of the future may bear little resemblance to those we deal with today. One of the fastest growing ad agencies in North America is a Canadian shop, Padulo Integrated, which is actually seven businesses under one roof. According to an April 10, 1993 article in the *Financial Times of Canada*, "This hot shop has rocketed up 48 percent a year in compound annual sales growth since 1985." The *Times* continues, "Rick Padulo's secret is a relationship with the client that is so synergistic that they really do become a partnership." According to Rick Padulo, president/CEO, Padulo Integrated added approximately $18 million worth of business in the first three weeks of October 1994. His greatest growth has been experienced during a deep Canadian economic depression. Padulo is doing some things right. His views of the future may be worth close examination.

Padulo's Insights for Media Salespeople

First, know the agency's business. He's shocked at how few radio salespeople bother to ask basic questions about agency executive philosophies—even understand the client list. This information is usually available from the receptionist or any account executive, the public relations people, and the media director. Get to know everything you can about the agency. Ask questions like: "Who makes this agency run?" "What are the strengths of the agency?" "Where did the agency begin?" Then, ask more questions. Second, realize that the agencies that spend large amounts weekly in broadcast or cable gross rating points (GRP) may well be dinosaurs. Those mass-market media budgets will shortly be shifted to budgets that are much more fragmented and less GRP-driven.

The New Media Mix	Weekly Investment
Broadcast television	15%
Cable	15%
Radio	15%
Print	15%
Promotion	15%
Direct marketing/Database	25%

Note: Less than half the budget goes into media that generate GRPs. The traditional agency would invest 100% of the above budget in broadcast television, generate loads of GRPs, gross $30,000 a week, and feel they performed good client service. Those agencies and stations that become involved with ongoing promotions, direct marketing, and database techniques will be winners in the twenty-first century.

Future Media Budgets

Rick Padulo believes that general media budgets will be slashed by about 50% from the levels of the mid-1990s. He sees far more emphasis on results instead of CPP approaches. He believes in transactions as ongoing cycles—not as one-time events. For instance, once Zellers sells a pair of socks, they immediately point to the need for the next pair. As soon as the politician wins an election, it's time to start working on being reelected. Break the cycle and you will fall behind in the world of niche and one-to-one marketing. The focus now is not on share-of-market—rather it is on share-of-customer. The emphasis is on building increasing sales from a company's customer base.

What to Expect From Agencies and Advertisers

Since George P. Rowell created the first advertising agency in 1888, the agency business has made many revolutionary changes. Yet we've only seen the tip-of-the-iceberg. Interactive media and the information superhighway are the darlings of the agency community today. Many ad agencies focus on the glamour of the visual side of this revolution. Interactive television brought to the consumer by the telcos or cable television fiber-optic paths fascinate many agency executives. The agency people tend to gloss over interactive 800 numbers—the touch-tone phones that interact with listeners—and information kiosks at points-of-purchase.

Expect to form media partnerships with progressive ad agencies. This means participating in one-to-one customer research with the agency research team. Envision the concept of continuous sales and marketing efforts for a particular retail or package-goods client. Learn the purchasing cycles of core consumers. This is the opposite of "being part of the media buy for a four-week ad schedule." Companies are looking at niche marketing opportunities. For example, Kraft is no longer having their agencies focus on the mass markets—people who purchase $200 to $300 of a particular Kraft product per year are no longer media targets. Their marketing focus is on the 20% core consumer who devours $600 or more of a particular Kraft brand. The marketing objective is not share-of-market, it's share-of-customer. Mass media budgets are being slashed. The budget is directed instead toward interactive media and direct marketing. To receive Kraft business, radio executives have to demonstrate how radio is interactive and can offer one-to-one vehicles like the KYS Connection described in Chapter 5.

The C.Q. Factor

Since the introduction of radio, the media business has been a business of intense competition and change. Now, the pace is accelerating. As Rick Padulo says, "We now live in a world where the level of your Change Quotient (C.Q.) is as important as the level of your Intelligence Quotient (I.Q.)."

17

The Importance of Building Partnerships

A Dream Come True

One of a book lover's greatest fantasies is to have access to a bookstore that carries a zillion books—and stays open until 12:00 AM. Books Unlimited would be a descriptive name for this store. Given an ample market from which to draw, one would think that business would soon be booming. The product would fill a niche and the product line would be diverse enough to attract a substantial number of readers within that niche. So far, the marketing logic is impeccable. The bookstore is attractively merchandised. Finding your desired section is uncomplicated and quick. The aisles are carpeted for comfort and to absorb noise. Benches are strategically placed in every aisle so that you can be comfortable while you leisurely examine the dust covers of books you may want to spend quiet evenings savoring.

The Promise, Then Apathy

Like most bookstores today, Books Unlimited will order books for you without additional charge. Recently, you purchased a book from this store. This self-improvement book really lit a fire within you. Your creativity and production levels have sky-rocketed. In 12 days, your best friend is going to have a birthday. You're excited about sharing the motivating information you've just learned with your friend. You decide to give her a copy of the book *Generating Wealth*. You return to Books Unlimited but find that they have sold their last copy of this inspiring book. You remember that they will order books, so you ask them to order a copy for you. "It'll take about a week to ten days to arrive," the pleasant clerk explains. You're elated because you'll have the book in time for your friend's birthday. Ten days pass. You haven't heard anything from the nice people at Books Unlimited.

You're very understanding but becoming a little uncomfortable. "Oh well, there's still two more days before I need the book," you tell yourself.

On the eleventh day, you admit that you're actually anxious about your friend's book arriving in time for her birthday luncheon tomorrow. You stop in at the store to check on your book's progress. The clerk checks his computer and reports that your book hasn't arrived. Before you can explain that *time* is of the essence, the clerk turns away to attend to other matters. You're beginning to feel like a sailor adrift in a dinghy in the middle of the ocean with a storm approaching. You decide to check the Get-Rich-In-24-Hours section of the bookstore where you found your copy of *Generating Wealth*. Within nine seconds, you discover two copies of the coveted publication. You're a little frustrated but grateful that you finally have your friend's present in your hands. You purchase one copy. Then, on your way out of the store, you stop by the "service" desk to inform the clerk that you have your book and that they can take your name from the special order list. A very disinterested clerk looks at you, makes no move to retrieve the order list on his computer, and says without enthusiasm, "Oh, okay." He then turns to answer a question from a fellow employee, totally forgetting your existence. "Oh well," you think to yourself, "I've tried to be a conscientious customer."

It's Not My Fault

The next evening about 9:30, your telephone rings. As you answer, you hear a female voice telling you that she is calling from Books Unlimited to let you know that the book you ordered has arrived. Since this is the *evening* of your friend's birthday *and* you've already attempted to notify the service desk representative that you have purchased a copy of the much-desired book, you're beginning to feel a little frustrated by the lack of communication within the bookstore. Your first thought is to ask yourself how you would feel if you hadn't decided to check the bookshelves yourself so that you could purchase your friend's present in time for her birthday. Your mental teakettle is beginning to produce steam as you remember the apathy from the "service" desk attendant when you tried to tell him to cancel your order. When you describe all this to the person on the phone, she quickly tells you it is not her fault. As you open your mouth to tell her you agree that it is not her fault, you hear the click of the receiver as she hangs up on you.

The Marriage Weakens

You remember the months that you anticipated the opening of Books Unlimited. You anxiously monitored the construction crew's progress as the building materialized. You often pictured yourself leisurely walking down aisle after aisle filled with books you would enjoy reading season after season. The other bookstores in your town never seemed to offer enough selections in the categories that interest you. The name of this new bookstore—Books Unlimited—assures you that *this* store will be different.

Now, only a few months later, following lackluster service time after time, the image has been tarnished. Sure, they have a larger selection of books than any store in town, but you don't feel particularly happy about visiting the store now. The excitement you felt when the store opened reminded you of the elation you felt when you discovered a new friend. Somehow, you feel betrayed. It's such a beautiful building. There are so many great books on its shelves, yet you're just not anxious to visit. When you *tell your friends* about your negative experiences at Books Unlimited, they have similar stories to relate about poor service at the store.

Oh, well, it is disappointing, but life goes on. You have other things to think about right now. Books Unlimited fades into the background of your thoughts—but just as it begins to disappear, you make a mental note to check that bookstore on First Street the next time you want a book. You decide to remind your friends about how warm and friendly the people are at that store. With all the excitement of the book giant opening its doors, you'd forgotten how pleasant it was to visit that rather small but intimate bookstore.

Trouble in Paradise

One of the market's most successful restaurant owners had been a loyal advertiser with Beach 103.1 since the station began offering ad packages that included marketing surveys. Rich, the owner, is a restaurant consultant as well as an owner. His restaurant, The Village Inn, is located in a coastal resort market not far from several large cities. When Beach 103.1's salespeople began to offer 26-week ad plans in combination with customized surveys just for his restaurant's customers, Rich was delighted.

The first 26-week program was considered a success. Advertiser and radio station personnel were pleased with the results. Rich and his sales rep, Edie, learned a lot about his customers' likes and dislikes. Rich and Edie were able to work together to develop several very effective ad campaigns for The Village Inn. Business flourished. During the 24th week of his ad program, Rich signed a renewal for another 26 weeks. The new program included an on-air dinner giveaway Monday through Friday as well as the ongoing customer survey.

Four weeks later, Rich called the radio station and left a message for Edie saying that he wanted to cancel the remainder of his contract. When Edie returned to the station after meeting with Rich, she was met at the door by her sales manager, the program director, and the general manager. "What's wrong?" they asked in unison. After talking with him, Edie reported that Rich was concerned because he felt people weren't redeeming the dinner certificates. He therefore concluded that his advertising *wasn't* working.

Go to the Source

Instead of second-guessing listeners, Edie went to the source and asked the contest winners: (1) Did they receive their certificates? (2) Had they used the certificates? (3) Why or why not? (4) If they *had* redeemed them, did they enjoy their experience? Exhibit 17.1 is a copy of the questionnaire Edie used when she called The Village Inn's contest winners.

Exhibit 17.1

Hi. My name is Edie. I'm with Beach 103.1 radio. Recently, _____ was a contest winner on our station and won a dinner for two to The Village Inn. Is this _____ ? (Is _____ there?)

We're taking a survey of contest winners to see if they liked our prizes. I need to ask you a couple of questions. Will you help me?

    If Yes, go to question 1.   If No, thank them politely and hang up.

1. Did you use your dinner certificate?
    If Yes, circle Yes and go to question 2.
    If No, ask "Why?"_____ Go to question 5.
2. Did you enjoy your dinner?
    If Yes, circle Yes and go to question 3.
    If No, ask "Why?"_____ Go to question 3.
3. How was the service?
    If answered positively, ask for details: _____
       Go to question 4.
    If answered negatively, ask for details: _____
       Go to question 4.
4. Would you recommend The Village Inn to a friend?
    If Yes, circle Yes and go to question 5.
    If No, ask for details: _____ Go to question 5.
5. Do you feel you need more than 2 weeks to redeem a dinner certificate?
    If Yes, ask how long: _____ Go to question 6.
    If No, go to question 6.
6. Did the amount of the gift certificate cover the cost of your dinner?
        Yes   No   (circle one)

Thanks for your help . . . and thanks for listening to Beach 103.1.

Defining the Universe

At the time of the presentation for the second 26-week advertising contract, Edie explained to Rich that he could expect an average of 3 to 5 winners per week, depending on the level of difficulty of the trivia question asked. Because the dinner giveaway had been on the air for 4 weeks, there were only 18 winners to contact, or attempt to contact. The telephone survey revealed many insights (see Exhibit 17.2).

Exhibit 17.2

The Village Inn — Beach 103.1

Contest Winner Survey

Survey conducted August 14 to 15, 199_
Attempted contact with 18 winners to date
Total contacted = 15

1. Did you use your dinner certificate?
   - Yes ------ 53%
   - No ------- 47%

2. Did you enjoy your dinner?
   - Yes ------ 88%
   - No ------- 12%

3. How was the service?
   - Good ----- 100%
   - Not good ---- 0%

4. Would you recommend The Village Inn to a friend?
   - Yes ------ 100%
   - No -------- 0%

5. Do you feel you need more than 2 weeks to redeem a dinner certificate?
   - 30 days ---- 63%
   - 2+ weeks --- 25%
   - 2 weeks ---- 12%

6. Did the amount of the gift certificate cover the cost of your dinner?
   - No ------- 62%
   - Yes ------ 38%

Survey Results

1. Did You Use Your Dinner Certificate? As the survey showed, 53% of the contest winners who were contacted redeemed their certificates. According to statistics for redeeming vouchers, this return ratio was phenomenal. This was an important component to point out to Rich. Because 3 of the contest winners could not be reached by telephone, this changed the universe of the survey to 15. Of the total certificates used, 53% translates to 8 couples. Now it's becoming clearer why Rich felt that "no one" was redeeming his certificates. Let's examine the 47% (i.e., 7 winners) who didn't use their vouchers. (The following percentages are rounded.)

The poll revealed that 29% (i.e., 2 winners) of those contacted—who didn't utilize their prize—explained that the certificate expired before they used it. Another 29% stated that they hadn't used them yet. One couple (i.e., 14%) reported that they had attempted to use their certificate at the restaurant even though it had expired, but they were refused. An additional 29% (i.e., 2) said that they didn't receive their dinner certificate. The sales rep double-checked the address listed for these contest winners and assured them that new certificates would be mailed.

As a result of this survey, radio station management instituted a new procedure, which included calling contest winners within 5 days of a certificate mailing, to make sure that the listeners had received their prizes. The radio station's management team and Edie were worried that Rich would be very upset when he heard that some winners weren't receiving their prize vouchers. Instead of being upset, however, Rich was reassured when he heard about the new follow-up policy.

2. Did You Enjoy Your Dinner? Overwhelmingly, 7 out of 8 people confirmed that they enjoyed their dinner at The Village Inn. The 12% who reported dissatisfaction with the meal translates to 1 contest winner. Since a customer who experiences unsatisfactory service tells approximately 8 people, it's very important to understand *why* someone had a negative encounter. In this instance, the customer felt that the veal parmesan was too salty.

3. How Was the Service? Rich was, of course, pleased with the excellent rating for his service. He had already told Edie that customers frequently gave his employees high marks for their service. It was gratifying to see it confirmed with Beach 103.1 listeners.

4. Would You Recommend The Village Inn to a Friend? With recipients reporting that they would recommend Rich's restaurant to a friend, there wasn't any other information to pursue in this category. Under different circumstances, you might ask your listeners to rate a client's restaurant based on numbers (e.g., 1 to 4) or stars (e.g., four-star restaurant being the best in your market).

5. Do You Feel You Need More Than Two Weeks to Redeem a Dinner Certificate? Almost two-thirds of those questioned gave many reasons why they needed up to 30 days to redeem their dinner certificates. The bottom line was that two weeks just did not give most people enough time to coordinate an evening out. At the time the contract was signed, Edie tried to persuade Rich to give the dinner certificates a life span of 30 days. Now, the customers were confirming the need for the 30-day redemption period. The client agreed to extend the time limit for dinner certificates to 30 days.

6. Did the Amount of the Gift Certificate Cover the Cost of Your Dinner? During the renewal presentation, Edie had explained to Rich that most listeners would spend more than the amount of the dinner giveaway certificates. Indeed, when questioned, the majority of respondents verified that they had spent between $12 and $20 more than the certificate provided. There were no reports that anyone was disturbed by this fact. Such information is very valuable when presenting a similar offer to other business decision makers.

The Bottom Line

When Rich heard that Beach 103.1 was willing to call listeners who had won dinner certificates and ask them why they weren't being redeemed, he was surprised. He was also privately impressed, as he confided to Edie, after seeing the results of the impromptu telephone survey. As a result of the information gleaned from contest winners, fine-tuning was performed on the promotion by Rich and Edie, employees at the restaurant, and employees at the radio station. Beach 103.1 managers held a full staff meeting to discuss what had transpired with The Village Inn account and how it affected *all* of them.

When the story was told to the on-air people, it gave them a better understanding of the information that was needed from them regarding *each* contest winner. The acceleration of events increases dramatically in the studio when an announcer is running a contest, taping call-ins, playing music, airing commercials, and attempting to get a winner's name and address all within what seems like a few seconds. It's no wonder that information can fall through the cracks during this high-speed juggling act. Understanding clearly what happens to the contest-winner data and how critical that information is to the success of a promotion—and the continued business of an advertiser—can be a valuable insight for announcers.

Office employees who are responsible for mailing prize certificates gained a new perspective on the importance of accurate information and the consequences of misinformation. Salespeople perceive all too clearly the end result of a weak link in a communications chain. A presumably happy client who calls the radio station in a lather and tells the *receptionist* to cancel his contract must be in the top three of a sales rep's Top-Ten Nightmares List. Bringing all the station employees together, however, helps *each one* see how vital their input is to the entire organization. This time an advertising contract was saved, a very valuable client saw how a marketing partnership was reinforced, and the infrastructure of a promotion was improved.

Delivering More Than You Promised

Some managers and salespeople may be outraged that a radio station *gave away* a survey to a client. "After all," they may think, "the client didn't even *ask* for the study." Any sales veteran knows that it's possible to make the sale but lose the customer. The survey conducted and the report the Beach 103.1 salesperson did required about three hours of her time. Edie never promised Rich that the station would poll the contest winners. However, when you consider how much time it takes to *acquire* a 26-week advertiser, three hours spent making telephone calls and then typing a report seems like a small investment. Studies reveal that radio stations spend approximately six times more to get new clients than they do to keep old ones. The fact that Rich was a long-term client makes him worth ten times the price of an advertiser who makes a single buy.

Handling Complaints

When handling complaints, you need to know two things: (1) What is the situation? (2) What does the client want—or how does he or she want it to be? Rich, the owner of The Village Inn, felt that his advertising wasn't working. In essence, he described the situation. When Edie questioned Rich, she found that he wasn't merely interested in canceling his contract as she had first believed; he simply wanted the advertising to be effective. At Beach 103.1, Edie had the autonomy to make on-the-spot decisions to fix a client's problem. She, therefore, suggested the poll and she invested *her* time in conducting the study. Advertiser surveys reveal that two-thirds of advertising clients who don't give you repeat business state that it's because of an attitude of indifference from their sales rep or other station employees. If you can resolve a client's problem *immediately*, almost 95% of advertisers will buy from you again.

Client Retention

Radio salespeople are often told that the most important aspect of their job is to gain advertisers for the station and then, to *keep* them. But, what does it take to gain a new advertiser? How many cold phone calls do you have to make before you even find a good prospect? How many visits do you have to make to a prospect's business before they become an advertiser with you? How much gasoline does your automobile have to consume? How much time do you have to spend preparing for your meetings with a prospect before he finally says, "Yes"? How much time do you have to spend analyzing each visit and trying to find ways your radio station fills the needs *you* believe you've uncovered? How much emotional energy do you need to expend trying to understand the refusals you're hearing? How much time do you need to spend correcting the misconceptions *you* had about a prospect's needs? Gaining clients—and then keeping them—is expensive for salespeople as well as the radio station. Client retention, therefore, is of the utmost importance. Think about the clients you have now. Ask yourself *every day* how you can be of greater help to them.

Analyzing Your Best

Think about your best advertising clients, that is, the ones with whom you have the most rewarding relationships. Examine those relationships. What makes them work? How do you *treat* each other? How do you *feel* about each other? Do you admire each other? Do you find yourself assisting, supporting, and advising any, or all, of your best clients? Have you ever stepped behind the counter and become a for-the-moment employee, giving your assistance for a few minutes when your client was short-staffed or the store had a rush of customers? Are you treated like an *insider* when you visit your advertiser? Are you included in the group when employees and ownership tell customer stories? Do you find yourself being supportive of your best clients? When they have a downturn in business or they've just lost their best salesperson, are you genuinely supportive? Do you search your mental employee bank to see if you may be able to suggest someone you've heard is looking for a similar position? Do you find yourself thinking about your client's organization and examining ideas you could suggest that might improve that client's business?

If you're analyzing your account list and considering which clients you would count as your *best* advertisers, ask yourself this question: "When I've endured a really stressful sales call and faced rejection, which clients come to mind as the *next* call to make because they accept me, respect me, and even forgive my mistakes while they buy from me time after time?" These are the advertisers that usually comprise your Best List.

Taking Your Time

When you think about sales calls to these Best Clients, it's likely that you spend more time during each call on their businesses. You probably also invest more time preparing for calls to their businesses. As you attempt to add more advertisers to your Best List, you may question whether you can afford to spend as much time with each new advertiser as you do with your Best Clients. Wouldn't you rather have more Best Clients than those advertisers who buy from you once or twice a year? Those fair-weather advertisers actually take a lot of your time because you seem to present idea after idea, or promotion after promotion, before they make a one-week buy. Examine your level of service to your Best Clients. Then, scrutinize your service to advertisers you would like to add to your Best List. The salesperson self-evaluation questionnaire shown in Exhibit 17.3 may help you evaluate the quality of your service to clients.

Exhibit 17.3

Client Satisfaction Questionnaire

Business name: _____   Advertiser's name: _____

Today's date: _____   Date of last evaluation: _____

Have I asked this client:

1. What would make you a satisfied client of our radio station? ☐ Yes ☐ No
   Answer: _____

2. What is good service from a radio sales rep? _____
   _____

3. How soon would you like to be contacted once your schedule begins airing?
   _____

4. Has your contact with other employees of our radio station been satisfactory?
   ☐ Yes ☐ No
   Why? _____

5. Are there any services our radio station could provide for you that you're not receiving from us now? _____

6. Have you received good value for your investment with our radio station?
   ☐ Yes ☐ No
   Why? _____

Withholding Satisfaction

Number six may be a difficult question to ask a client. Most salespeople are afraid of the answer. Think about servers in a restaurant. They *must* ask each customer, "Is everything OK?" Checking with the customer to see if their meal is satisfactory is putting the server and the restaurant on the line. Ironically, sales reps are keeping themselves from receiving positive feedback when they avoid asking this question of their advertisers.

Admittedly, some clients fear giving a positive answer because they believe it just opens the door for their sales rep to ask them to buy more advertising. Take chances. Challenge yourself to ask question number six. You'll be amazed by the amount of positive feedback, which will inspire you to find more ways to satisfy your clients.

Being a Good Partner

Seeing Through Their Eyes

Walk through your client's store, or office, looking at everyone and everything there as if you were the owner. What is important to you? What concerns you? What gives you the most pride? While you're very proud of your company, do you feel an obligation to all of the people who work there? Do you feel an obligation to make the organization as successful as possible—not just for yourself, but also for your family and for all of your employees who depend on *their* jobs to give *them* security for **their** families? Do you look at the inventory and equipment that is necessary to run this operation and realize that a great deal of money and human energy has been invested to make it a success? Do you feel a sense of pride for being able to bring all this—people and accoutrements—together to create an organization that provides a valuable service to others or provides materials for others? Do you think of the ways your client wants her parents, her mate, and her family to be proud of her?

As you stroll through this client's business, do you also feel the anxiety she feels when she considers all the decisions that must be made on a daily basis, decisions that may affect the prosperity or failure of this venture? Do you sense the delicate balance between pricing, sales, and cash flow? Finally, add the human factor of all the politics that permeate the air anytime two or more human beings interact, whether those humans are employees or customers. Now, ask yourself one more question. How can I, and my radio station, help this person?

Developing Intimacy

Don't misunderstand. We're not talking about knowing your clients in the *biblical* sense. We're simply suggesting that you get to know your clients as well as you know your closest friends. It takes a lot of dedication and perseverance to learn about your advertisers' innermost beliefs, viewpoints, and feelings. Performing Niche Marketing Analyses will help you *begin* this process. Then, when you think you know your clients as well as anyone can, it's time to begin the process all over again. As you know, everything changes, *constantly*. Clients' circumstances literally change from hour to hour. You'll need to keep abreast of those changes if you intend to remain competitive.

Reach Out and Touch

Call your clients and thank them for doing business with you. Some stations mail thank-you notes to advertisers. That's a positive step. A personal phone call may have more impact, however. Call your advertisers and thank them for paying their bills consistently—and on time. You'll be surprised how much this one small gesture can mean to clients. Call your advertisers often. Call at least once a week.

Invite them to your radio station to speak with your salespeople about *their* perception of customer service. Suggest that some businesses have job swaps with your station. For example, your client's marketing director or advertising buyer may exchange places with your program director. Consider the insights that may be gained. Propose that some of your clients participate in a focus group discussion.

Write letters or notes to your advertisers. Some salespeople are very diligent about sending notes to their clients. Even the most diligent can increase their communications with advertisers. Send magazine articles about their type of business, about new marketing trends, about training employees, about customer service, about their hobbies, and even about raising children—if they have any or are contemplating creating some.

When you call, have a purpose for the call. We're not suggesting that you call and ask, "Planning on doing some advertising this week?" Just like your closest friends, you call to find out how a particular project is progressing or to share a piece of information that may be beneficial to their businesses or to them personally. Don't be shy about staying in contact with your advertisers. Convince them that you are truly interested in their success. Think of *new* ways you can stay close to your clients and prospects. Begin now.

Honeymoon Sales Calls

One sales rep went to Europe on her honeymoon. She knew she would be gone for three weeks. She also knew that was a long time to leave her clients alone with her competitors. While she was adventuring through Europe, she regularly sent postcards to her advertisers. She included her highest-billing clients, advertisers with whom she was trying to develop relationships, and some advertisers that had never purchased ads from her. The notes on the postcards were short, but they sent a powerful message to her clients and prospects. She wrote simple notes similar to the ones she would send to her closest friends. Surprisingly, it took little time away from the activities of her European adventure. While the newlyweds were sightseeing, the bride often took snapshots of businesses that were like the businesses of her clients.

When she returned home, she had extra photos printed for her clients. These photos she presented in person on the very first call she made at each client's office. She related stories about how different, or similar, the European businesses were to that client's business. Her clients were surprised—and delighted. She reinforced her intimacy with her clients and made great strides with her prospects. Was it worth the extra effort? She answers confidently, "Yes!"

Show Your Appreciation

Everyone likes to feel appreciated and made to feel important. Alison, a sales rep at an AM daytimer was trying to convince a significant advertiser of another radio station to become a long-term client of her station. David, the owner of Ford Jewelers, had been making periodic buys with Alison's station. He told her how impressed he had been when her general manager visited his store the previous week to thank him for his business. David also confided that during the 20 years he had been on the top-ten list of her competitor's advertisers, neither their general manager nor their sales manager had ever had any contact with him, or ever visited his business. Yes, David did become a major advertiser with Alison's station. He also became an excellent source for referrals.

Nurturing Partnerships

My Partner, Evelyn

I, Ashley, think I'm a typical customer at my hairdresser's salon. I don't expect too much when I make an appointment for a haircut. I just want the haircut to make me a couple of inches taller, increase my bust size by one letter, and take 10 years of wear-and-tear off my face. Intellectually, I know this is unlikely to happen but when I look in the mirror after Evelyn, my hair stylist, pronounces me beautiful—her term for *that's all I can do*—I'm just a little surprised that it is still me reflected in the looking glass. Given these modest expectations of Evelyn, it's truly a miracle that I've been her loyal customer for seven years.

It's not that I haven't considered divorcing Evelyn and WHOOPDEE-DO! (her salon) over the years. There have been times when my frustration level has risen to about a 6.3 on the Richter scale. There are also times when I don't feel that Evelyn is really listening to me. Sometimes she's distracted by her personal life. Once in a while my appointment follows someone who has already drained her energy to a place about six feet below sea level. Then, there are the times when *I've* been rushing from client to client and arrive at Evelyn's shop breathless—but on time. That's just the moment when she makes me wait for 15 minutes while she takes care of a customer who rarely makes appointments one day in advance. So, why am I so loyal?

Splitting Hairs

Evelyn is very talented. She has created many flattering haircuts for me. Most of the time, she listens attentively when I attempt to describe a hairstyle I've seen on a well-known person. While *I'm* absolutely convinced that this hairstyle will fulfill all my expectations for the perfect cut, she patiently explains that this celebrity has eight hairs for every one of mine. Evelyn often suggests a similar style that will work better for my type of hair. Somehow, I'm not offended or frustrated. I'm not an amateur at the hair salon experience either. I've been to many salons around the world. I've been told—in many languages—how gorgeous my new haircut looks on me. Like most people, I'm not immune to flattery, but somehow when Evelyn tells me that my hair looks terrific, I feel a little better about myself. There are also the times when *I* arrive 15 minutes late for my appointment. Evelyn always reacts graciously even though I know her stress level must increase because she now has to satisfy my needs in less time than she usually requires in order to be ready for her next customer.

Like any partnership, ours has moments when one partner falls short of the promises *assumed* in an (often unwritten and unspoken) contract between us. Overall, however, we have a 50–50 partnership. We also understand that in order to maintain that balance each must sometimes give 60% or 80% while our partner is only able to give 20% or 40%. Without appearing obvious, Evelyn works very hard to keep our partnership balanced. And, since she is sincere in her efforts, I am inspired to be the best customer I can be. But is that enough to motivate me to continue our business relationship?

One of the Family

It's just that she makes me feel like I'm one of her best friends—or even family. I trust her. It seems like we've shared so much of our lives with each other even though we don't see one another outside of our monthly rendezvous at her shop. Ah, but those hours spent in her shop—while she performs miracles on my hair—have been some of the most laugh-filled, tearful, camaraderie-filled, and sisterhood bonding that I have ever experienced as we share our life-events with each other. I know that other talented stylists are in my market. It's just that I'm content in my partnership with Evelyn. Whenever I contemplate changing hairdressers, I think of all the times Evelyn has helped me *above and beyond* the call of duty. I recall one occasion last year.

Being There

At five o'clock on a Tuesday afternoon, I learned that I needed to be in another state on Wednesday morning for a business meeting. I also had an appointment with Evelyn for our monthly meeting the next day. Since I wanted to make a good impression on my prospective client, I didn't want to appear at the meeting looking like I was the stand-in for Lassie. I called Evelyn to explain why I couldn't keep my appointment for the next day. Without taking a breath, Evelyn asked if I could drive to her shop right then. She told me she would wait for me and even have a glass of wine poured. Even though she was very gracious, I knew it was an imposition; but I also knew that I would feel more confident at the meeting the next day if I felt I looked my best. At that moment, I felt like Evelyn was my sister and best friend. I knew she understood my situation but more than that, she cared.

Why am I so loyal? Do you think that the next time my haircut didn't turn out just as I had envisioned or that Evelyn kept me waiting for 15 minutes because she had a fussy client that *I* didn't understand *her* dilemma? I care about her because she cares about me. In an era when it costs more to gain a new customer than it does to keep a current customer, Evelyn has earned my admiration, devotion, and loyalty. My customer partnership has paid dividends for Evelyn also. I have sent my friends, my family, and visiting relatives to Evelyn. Great customer service pays and pays and pays.

Generosity Encourages Loyalty

Examine what Evelyn does to earn her customers' loyalty year after year. She begins her relationships with an attitude of generosity. Evelyn knows that one of the greatest hairstyle aggravations for females who have bangs is the fact that they often grow into your eyes before you really need another haircut. She also understands what a nuisance it is to the person who spends what seems like most of her day pushing or blowing her bangs out of her eyes over and over again. This annoyance can be very distracting and could cause one to feel negatively toward the person who *cut* the bangs in the first place. To compensate for the exceptionally fast growth of bangs, Evelyn invites those customers who wear bangs to return to her shop at any time between appointments for a free trim. Not every customer who has bangs in her hairstyle takes Evelyn up on her offer, but even the ones who forgo the offer feel better just knowing that Evelyn is sincere and the option is available.

Evelyn has also been generous with customers who ask for a particular style, seem happy with their cut while they're in her chair, but later decide that it isn't exactly what they had pictured or believe that it needs additional trimming. Evelyn can sense a customer's satisfaction rating sinking off the scale. She has often invited the customer to return to her shop as soon as she can schedule her, and she makes *every* attempt to schedule her as soon as humanly possible. Evelyn gives all of her attention to a customer like this during her return visit. When the customer says she is satisfied, Evelyn quietly tells her that there will be no charge. As you can imagine, this way of doing business could become *very* expensive for Evelyn. Because of her attitude of generosity, however, she never focuses on what she could lose but on what she can gain. Thus, customer loyalty is nourished.

Balance and Listening

Just as in your relationships with your mate, your children, your parents, or your friends, your business relationships need balance. Balance doesn't mean that every transaction between you and your client must be equal. It does mean that *overall*, each person must *perceive* that their needs are met about 50% of the time. In actuality, the symmetry may not be 50–50 but when each of you perceives a balance, the relationship is on course. "How can my needs be met half of the time? I'm always doing things for my advertisers. What do they do for me—besides sign a contract?" a salesperson may ask. Some of the ways your clients can give to the relationship include keeping their word with you, having their copy ready when they promise, paying their bills on time, and keeping their appointments with you.

Now, list all the ways you can contribute balance to client relationships. Consider your clients' actions. Do you mirror them? Do you keep your word? When a client tries to contact you, do you return phone calls as soon as possible? Do you pick up copy when you promise? If you say that you'll be in on Friday morning to pick up a check, do you follow through? Are you on time for appointments, or do you have your "most popular" list of reasons why you're often late for meetings? Do you listen—that means really listen, listen, and listen—to your clients? Do you listen to your advertiser's words and then listen *between* the words? Sometimes your contribution may be to understand what clients are not saying as well as what they are saying.

Before you burst with frustration because you feel that you didn't agree to become a psychologist for your advertisers, consider what you do in your relationships with family and friends. You listen to what they're saying *and* to what they *aren't* saying so you can contribute to your relationships as much as possible. Why do you feel such a need to contribute with family and friends? You realize that sometime in the future you may need *their* complete attention in order for them to help you. Listen intently to your clients. Sometime in the future you may also need *their* undivided attention when you're offering marketing or advertising advice.

Trust

Office Depot has decided to trust its customers. For the convenience of patrons, Office Depot locations have self-help copy centers. Anyone can use their plain-paper copiers. Customers are provided spacious worktables, which hold paper clips, staplers, paper punches, and even a machine that inserts comb binding into a pile of papers to create a booklet. Except for the copier, there is no charge for using these machines or tools; plus, copies are usually priced lower than anywhere else in the market. When you're finished making copies, a blue printed sheet of paper is conveniently provided next to the copier. You can easily list the number of sheets you've copied. You may also calculate the cost, but you don't even have to do that because the clerk at the checkout will do it for you, if you prefer. On checkout, no one is ever questioned about the number of copies reported on the blue slip; *no* copies are ever counted.

Of course, Office Depot knows that some people will underreport the actual number of copies made. But they also know that what they lose in extra copies is more than made up for by other business these customers give to the store. Like milk and bread in a grocery store, the copy center is strategically placed in a far corner at the back of the store. Customers must walk down wide aisles filled with thousands of temptations for businesspeople and students. Office Depot is building top-of-mind awareness of their inventory each time a consumer walks down their aisles. The ultra-economic copy center also provides daily traffic composed of customers who make impulse buys. These impulse buys are augmented by more substantial purchases of anything from computers to office furniture. Originally, small-business owners and students were the loyal devotees of Office Depot stores, but now even mid-sized businesses can be counted among their customers. People like to be trusted. It's an essential component of any successful relationship. If independently questioned, would your clients answer that you (and your radio station) trust them?

Flexibility

One radio station's largest advertiser was the owner of an optical shop that had two locations. From time to time, the owner would slip behind in his payments to the station. The salesperson who called on the optometrist would go into the shop to ask for a check. He was never refused. Even though cash flow was sometimes erratic for the advertiser, he always found a way to pay a large portion of his bill or made arrangements to give postdated checks to his sales rep. Being late with a payment or paying portions of his bill always embarrassed the client even though his sales rep was understanding and very discreet. Radio station ownership had originally laid the collection plans for this client. Ten years passed.

Reliability

The Optical Shop remained the station's largest advertiser. Like waves breaking on the shore at low tide and high tide, during lean times the client's bill grew, and during fruitful times the bill waned. Every year when contract renewal time arrived, the salesperson and the owner discussed plans for the next year. Each year the radio rep included a rate increase. The increase equaled but never exceeded the rate of inflation. When rate negotiations were being conducted, the sales rep was able to justify the increase. The advertising campaigns were working well for the optometrist, the rate increases seemed justified, and the client remembered how well the radio station worked with him to develop new ideas. He could trust his rep and he counted on his rep's discretion with his bills. In return, he always kept his word. The shop owner knew that partners depend on each other—they don't waste time or energy wondering if the other person will *cover their back*.

New Owners Can Lose Partners

In the eleventh year of the optometrist's business association with the radio station, the station was sold. The new owner introduced new policies. When the new owner saw that the optometrist was not current with his payments, he demanded that The Optical Shop be taken off the air until payment-in-full was made to the station. The radio station transition occurred during a slow time for the optical client, so he was unable to pay all of his bill at one time. The new owners weren't interested in hearing any reasons why the client was unable to pay the entire bill immediately. They didn't check his impressive payment record. The new station owner looked at the situation as though the client had taken advantage of the previous owners. He was going to make sure this advertiser knew who was in control at his station. The owner of The Optical Shop made monthly payments on his account until it was paid. As usual, he kept his word. Unfortunately, the radio station lost its largest advertiser. The new owner didn't base his business relationships on trust and flexibility—or an outstanding track record.

Crystallizing Expectations

If you and your client partner have a clear picture of what each of you will receive from the association, you'll have less chance of disappointing each other in the future. Encourage your advertisers to be clear about their expectations of the relationship. The salesperson's self-evaluation form—the Client Satisfaction Questionnaire, Exhibit 17.3 in this chapter—will help you begin the process of crystallizing your clients' expectations. Questions from the Niche Marketing Analysis: Part I form in Chapter 4 will also aid in this exercise. Imagine looking at your relationship through your advertisers' eyes. Is it easy to be a customer of your radio station? Do your service expectations frustrate your clients or assist them?

252 *Clients as Partners and Looking Ahead*

Lost in a Fog of Vagueness

Oldies 105 was known throughout their market for a consumer show the radio station held every spring. At the time the show contract was signed, each advertiser was asked for a 50% deposit to secure booth space. The balance was due before set-up at the show. For many years, set-up day for the exhibitors was a nightmare. The people assembling the displays usually weren't the people who had signed the contract for the show. When people arrived to find their booth space and put their displays together, there were always arguments. Frequently, the set-up people had no knowledge of the fee due before they could unload their equipment. They also had no authority or ability to issue a check. You can imagine how charged the atmosphere became at the service entrance to the building. If a client's employees were fortunate, their manager had paid in advance; and they were admitted through the "pearly" gates to their exhibit space.

The next hurdle appeared when people tried to stake out their territory. It's amazing how human beings can transform into territorial cave dwellers when they are trying to claim a ten-by-ten foot space in an exhibition hall. Booth space was charted by drawing what appeared to be hieroglyphics on the floor with chalk. Each exhibitor seemed determined to use just one extra foot of their neighbors' space on each side of their booth. Verbal war clubs were often drawn. Add the process of finding adequate electrical outlets for each exhibitor, and the traffic jam at the service entrance because no one was given a time range for unloading their supplies. The endless set-up day claimed the patience of the sales staff as they tried to defend *their* clients' demands. The exhibitors were next on the mortality list. The sales manager, Jeff Rivers, was also a victim of frazzled nerves during this excruciatingly painful process because he was the floor supervisor. The mood was now set for the show. Could success be found anywhere in this universe?

Being Explicit

After many years and a lot of trial, and mostly error, Oldies 105's consumer show slowly became more organized. Communications with clients improved because communications within the radio station improved. When Jeff introduced the sales package for the approaching consumer show, he also showed the salespeople a sample contract, which contained all of the pertinent information needed to smooth the path for an orderly and peaceful set-up day at the exhibition hall. This year, the radio station would be hiring a drapery rental company to define each exhibitor's display space the day before exhibitors arrived with their supplies. The rental company employees would create each booth with curtains from a floor plan design based on the terms outlined on the signed contracts. Each contract would state the size of booth, its number on the show floor plan, the total price paid, the check number from the 50% deposit, the electrical requirements of the client, and the time range for the advertiser's employees to set up at the show.

Salespeople were required to phone the station at the time the contract was signed and confirm booth space. Times for delivering equipment were to be determined in advance based on booth location. The innermost booth spaces would be filled first. Then, booths were to be filled in descending order as one approached the service entrance. An iron-clad rule was established at the radio station. As sales manager, Jeff would not approve any contracts that didn't contain *all* of the necessary information about the show. Understanding the aspects of successful relationships, Jeff kept his word with his salespeople. The salespeople trusted that Jeff meant what he said. They delivered completed contracts.

Clarifying Details—A New Era

Ten days before the show, the station's general manager sent letters to all exhibitors. She explained where certain amenities could be located for the comfort of the exhibitors and described regulations to govern the safety of people using the building. Security procedures were outlined. In effect, she presented the rules and regulations as if the radio station's team was expecting guests. She made it clear that people's needs and comfort were of utmost importance. In addition, copies of this letter were posted in the exhibitors' lounge and near the service entrance during set-up. Advertisers were encouraged to phone the station if they had any questions or comments.

A Week Before the Show. Salespeople called clients who had not paid the remainder of their balance for the show. The reps tactfully reminded advertisers of their balance due, and explained to clients that their goal was to produce a very positive experience for everyone participating in the show. Appointments were made to pick up the exhibitors' checks. Salespeople also checked to see if clients had any last-minute questions during these calls. Often, sales reps volunteered to brief employees if there were any queries.

Three Days Before the Show. The sales manager's assistant made copies of all contracts for the show and presented them in a binder to Jeff. He made sure this binder accompanied him to the exhibition hall. If there were any questions regarding what was promised or purchased, ruffled feathers were more easily smoothed when Jeff produced the contract notebook and reviewed the signed contract with a possibly disgruntled exhibitor. Set-up day for the show calmed down approximately 99% from the previous year. Many more exhibitors volunteered positive comments about the exemplary organization of the show. Because the radio station valued and respected the input of their clients, a survey was conducted after the show. Exhibitors were asked how the show could be improved the following year. The client–radio station partnership grew stronger. Conducting business with Oldies 105 became easier and easier.

Is It Easy to Do Business with Your Radio Station?

Look at your billing statements through the eyes of a stranger to your industry. Do your invoices look like a mathematical maze? Do you describe an advertiser's expenditures in a language that only your billing department understands? Do your clients need a media dictionary and a manual for computer analysts to read your statements? Could a layperson (i.e., someone who doesn't buy advertising) read your invoices and understand them? Would an amateur know what your client had purchased? Would they readily grasp how much was owed—and when it was due? Are interest rates easy to read or relegated to small print at the bottom of the page or even on the back side of your bills? Some of the only tangible proof that an advertiser receives from his buy is a billing statement. Is yours *truly* user-friendly?

Testing and Learning

Test your billing department employees who are in contact with your advertisers. Give an example of a possible scenario where a client contacts them with a complaint or question. Ask how they would handle the situation. The ensuing discussion may give you a great opportunity to learn from your representatives and possibly to clarify employee misconceptions. Is the rest of your radio station user-friendly to advertisers? Are salespeople user-friendly to advertisers? Are production people user-friendly to your clients?

Appraise

Appraise everything your radio station does related to your advertisers. Survey your clients. Find out how they define customer service. Ask what constitutes a great partnership. Begin by using the forms in this book. Develop more questions. When you've gathered this information, analyze it. Decide where you can improve. If a part of your mission statement is to provide the best service your advertisers have ever received—and you strive to accomplish it—your radio station's growth and profitability will take care of itself.

18

The Future

Groupies

Growing numbers of radio stations are controlling multiple stations in given markets. Every group—if they're astute—will program and market to specific (noncompetitive) demographic niches. For example, three stations in a group may program to women 18–24, 25–34, and 35–44. Six or seven stations per market may be in each group. These groups will offer marketing strategies and sell marketing services (e.g., an ad campaign) to each station's clients. Full-service art departments will create station promotional materials. Graphics for advertisers will also be offered. Copywriting staffs—with writers who can solve clients' marketing and advertising problems—will be available. These services will create *new* revenue streams for radio stations.

FIGURE 18.1

Is There a Future for Individual Stations?

Those stations that choose to add niche marketing professionals—copywriters who are versed in niche selling and salespeople who are true niche marketing consultants—will survive in their chosen niche. The new breed of sales managers and salespeople will be skilled in *all* media. Last-minute copy, quick-fix sales packages, and poor long-term planning will be as obsolete as mass marketing. The ability to take a client's entire advertising budget and suggest multimedia solutions will ensure individual radio stations a place in the advertising budgets of the future.

Opportunities for Everyone

The techniques discussed in this book are equally available to radio's many competitors—cellular phones with information-on-demand; interactive cable and television linked to computers; database telco services for newspapers, shoppers, and individually addressed magazines. Radio is still in a unique position to capitalize on niched database information and one-to-one marketing research. Skilled, niche marketing radio people can succeed in spite of the explosion of competing media.

An *Advertising Age* Alert

As early as 1990, *Advertising Age* reported that the H. J. Heinz Company was ending its 36-year relationship with the giant ad agency Leo Burnett. The article quoted a Heinz spokeswoman: "There was a time when the one-size-fits-all worked for our brands. But our approach is changing from traditional to nontraditional marketing—targeted marketing and micromarketing. The broad full-scale advertising that Leo Burnett does so well is no longer the best approach for us." The Heinz move has ramifications for all media people who continue to engage in the business of mass marketing.

A Road Map

This book has been designed to act as a road map for those individuals and groups who are ready to take the next logical steps in media leadership. The ideas presented here are designed to help you break the 7% to 10% share that *all* radio stations traditionally acquire in a particular market. This revenue is going to drop as advertisers seek more definitive information about CPP—cost per prospect—and demand greater results from their former mass media partnerships. The traditional advertising pie will be sliced thinner, thanks in part to database marketing, which is now available to all your media competitors. Niching, database techniques, and one-to-one research can help radio stations rise above their media competition in an interactive environment. You don't need a sophisticated hardware system in small markets. You just need more information. A $60,000 hardware system from a company like Brite Voice may enhance your presence, however, in larger markets. If radio doesn't meet the challenge, newspapers will offer the service and gain the additional marketing revenues these new information services afford—not to mention the one-to-one sales opportunities offered by direct telco connections.

FutureMedia

In *PowerShift*, futurist Alvin Toffler (1990) describes five criteria for future media. As you read his descriptions think about *radio*. You'll see that radio—more than any other medium—*can* meet the communication demands of the future. According to Toffler, the five hallmarks of the media in our future are:

1. Interactivity
2. Mobility
3. Compatibility
4. Connectivity
5. Globalization

When describing the media of the future, Toffler and marketers like Peter Drucker rarely mention radio.

To change these perceptions of reality, radio needs to be *re*packaged as a new media. One way to repackage radio is to make it more interactive. Engineers tell us that by the year 2010 every space on the surface of the earth will have at least 500 media options available. Radio can be *imaged* as the portable medium of choice. In this new media scenario, radio stations that grow can no longer accept the position of being a *secondary* source of information and entertainment. The ultimate test, however, will be how radio operators meet the evolutionary demands of their heavy, heavy users—steady advertisers and core listeners. As the idea that radio meets these new consumer and advertiser demands is accepted, you can create niche marketing success stories.

Research Is Key

Don't be afraid to conduct your own research. Your market as well as your clients' customer base are always changing. To remain competitive, consider making surveys and customer contact a new way of life. Remember, professional researchers often use unskilled employees to conduct face-to-face and telephone interviews. The intelligence they gather has to be tabulated and analyzed, but this can be accomplished by the same unskilled workers who conduct the interviews. Or different people can tabulate and analyze the data. Sometimes, there are handwritten notations that must be deciphered. Add report writers to the list of people involved. This chain of work may pass through as many as 5 to 10 different people. Remember the *whispering game* you played with your friends when you were very young?

The Whispering Game

You and your friends stood in a circle. You whispered one short sentence to the friend on your right. She whispered that sentence to the person on her right. In turn, each person whispered the sentence he or she heard until the message returned to you. Except, it wasn't *anything* like the sentence you started! In fact, it probably wasn't recognizable. As human beings, we *interpret* what we *hear*, what we *read*, and what we *see*. We *rarely* repeat something verbatim. Most researchers will tell you that compiling research demands judgment calls. Those judgment calls may drastically affect your final research conclusions. If you truly want to become marketing partners with your clients, conduct research for them. Remember, you don't have to know all the answers, you just need some good questions. Successful niche marketing requires gathering many pieces of information over a period of time—a once-a-year survey is *not* enough. To remain competitive, consider making niche marketing research a new way of life at your radio station.

Another Perspective

There is one theme woven through every chapter of this book. Improve your communication abilities and you'll increase your business. Effective communication requires an unending quest to understand other people's perspective. An unexpected bonus of this crusade for better communication in the workplace will be better communication in your personal life as well. The benefits will multiply. Make your sales efforts *advertiser-driven* by gaining their perspective of business. Help your advertisers make their sales and marketing efforts *customer-driven* by learning their consumers' perspective—you'll both enjoy more success.

Their *Shoes*

Put yourself in the shoes of your advertisers one more time. You manage a business. You have many, many, many media reps calling on you. They all—more or less—tell you the same media story. They tell you how they're number one in some category. They tell you that they want you to advertise with them. They tell you about their package-of-the-hour. Some have funnier jokes than others, but they *all* seem to blur in your mind's eye because of their *sameness*. Except for one rep who comes in and asks questions about your business instead of telling you about the latest sales promotion-of-the-day. This rep actually tells you about marketing techniques to keep your customers coming back time after time. It's always a pleasure to see this salesperson because the rep gets you to think new thoughts and see your customers from a different perspective. This rep's advertising suggestions make a lot of sense, and the ideas seem to have been created just for your business.

Now, step back into your own radio shoes. Who do you think this businessperson is going to buy *more* advertising from—*more* consistently? Who do you believe gets higher rates from this advertiser? Who has easy access to this client? Wouldn't you rather *be* that rep than someone competing with her?

The Odds Are in Your Favor

You can win the niche marketing battles against all local media . . . including cable. You're entrenched. There are more radio stations in the world than any other medium. The odds are now in radio's favor. Looking for new business opportunities? On average, 1,700 new corporations are started every business day in the United States alone! Those that survive will concentrate their advertising and marketing efforts on specific niches. Your job is to deliver more than a format. Show your clients how to capitalize on your station's unique strengths. Demonstrate highly targeted radio formats tied to intelligent niche marketing plans, utilizing *your* radio station.

The New Breed of Radio Salespeople

Recruit people who are dedicated marketing people. Look for individuals who understand the value of *all* media but *choose* to work in radio. Give them autonomy with client advertising and marketing plans. Make them an integral part of your staff. Compensate them well. Position your station as the local niche marketing source. Your marketing staff can help your clients win *their* niche marketing battles.

Happy Niching
It's 2010, the picture of media's future drawn by the Herwegs has crystallized. You are one of the fortunate ones—you were prepared.

References

The Focus Group, by Jane Farley Templeton, Probus Publishing Company, Chicago, 1994

Making Niche Marketing Work, by Robert Linneman and John Stanton, McGraw-Hill, New York, 1994

Positioning: The Battle for Your Mind, by Al Ries and Jack Trout, Warner Books, New York, 1981

PowerShift, by Alvin Toffler, Bantam Books, New York, 1990

Radio Advertising's Missing Ingredient: The Optimum Effective Scheduling System, 2nd Ed., by Pierre Bouvard and Steve Marx, The National Association of Broadcasters, Washington, DC, 1994

Recruiting, Interviewing, Hiring, and Developing SUPERIOR SALESPEOPLE, by Godfrey W. and Ashley Page Herweg, available through The National Association of Broadcasters, Washington DC, 1994

Suggested Further Reading

Body Language, by Julius Fast, Pocket Books, New York, 1970

Bottom-Up Marketing, by Al Ries and Jack Trout, McGraw-Hill, New York, 1989

The Clustering of America, by Michael J. Weiss, Harper & Row, New York, 1988

The Complete Database Marketer: Tapping Your Customer Base to Maximize Sales and Increase Profits, by Arthur M. Hughes, Probus Publishing Company, Chicago, 1991

The Copy Workshop Workbook, by Bruce Bendinger, The Copy Workshop, Chicago, IL, 1993

Creating Money: Keys to Abundance, by Sanaya Roman and Duane Packer, H. J. Kramer, Inc., Tiburon, CA, 1988

Customers as Partners, by Chip R. Bell, Berrett-Koehler Publishers, San Francisco, 1994

Dare to Connect: Reaching Out in Romance, Friendship, and the Workplace, by Susan Jeffers, Fawcett Columbine, New York, 1992

Empowerment Takes More Than a Minute, by Ken Blanchard, John P. Carlos, and Alan Randolph, Berrett-Koehler Publishers, San Francisco, 1996

Feel the Fear and Do It Anyway: Reaching Out in Romance, Friendship, and the Workplace, by Susan Jeffers, Fawcett Columbine, New York, 1987

The Great Marketing Turnaround: The Age of the Individual—and How to Profit From It, by Stan Rapp and Tom Collins, The Penguin Group, New York, 1992

Hitting the Sweet Spot, by Lisa Fortini-Campbell, Bruce Bendinger Creative Communications, Inc., Chicago, 1992

How to Win Customers and Keep Them for Life, by Michael LeBoeuf, The Berkley Publishing Group, New York, 1987

The Intuitive Edge, by Philip Goldberg, Jeremy P. Tarcher, Inc., distributed by Houghton Mifflin Company, Boston, 1983

Listening: The Forgotten Skill, by Madelyn Burley-Allen, John Wiley & Sons, Inc., New York, 1982

Market Mapping: How to Use Revolutionary New Software to Find, Analyze, and Keep Customers, by Sunny Baker and Kim Baker, McGraw-Hill, New York, 1993

Marketing Warfare, by Al Ries and Jack Trout, Penguin Books, New York, 1986

Measuring Customer Satisfaction, by Richard F. Gerson, Crips Publications, Menlo Park, CA, 1993

Men Are From Mars, Women Are From Venus, by John Gray, HarperCollins Publishers, New York, 1992

The Mind, by Anthony Smith, The Viking Press, New York, 1984

Motivation in the Real World: The Art of Getting Extra Effort From Everyone—Including Yourself, by Saul W. Gellerman, A Dutton Book, published by the Penguin Group, New York, 1992

The One to One Future: Building Relationships One Customer at a Time, by Don Peppers and Martha Rogers, Currency-Doubleday, New York, 1993

The Power of Logical Thinking, by Marilyn vos Savant, St. Martin's Press, New York, 1996

Procrastination: Why You Do It, What to Do About It, by Jane B. Burka and Lenora M. Yuen, Addison-Wesley, Reading, MA, 1983

Real Power: Stages of Personal Power in Organizations, by Janet O. Hagberg, Winston Press, Minneapolis, MN, 1984

Selling the Story: The Layman's Guide to Collecting and Communicating Demographic Information, by William Dunn, American Demographics Books, Ithaca, NY, 1992

State of the Art Marketing Research, by A. B. Blankenship and George Edward Breen, NTC Publishing Group, Lincolnwood, IL, 1992

The 22 Immutable Laws of Marketing: Violate Them at Your Own Risk, by Al Ries and Jack Trout, HarperCollins Publishers, New York, 1993

Use Both Sides of Your Brain, by Tony Buzan, E. P. Dutton, New York, 1983

Winning Moves: The Body Language of Selling, by Ken Delmar, Warner Books, New York, 1984

You Just Don't Understand: Women and Men in Conversation, by Deborah Tannen, Ballantine Books, New York, 1990

After Words

How do you say thank you to the person who has been the most positive and loving influence in your life? How do you show your gratitude to someone who has been the most caring human you've ever encountered? Someone who had a wondrous love of life. Someone who dearly loved the radio industry. He did indeed want to jump out of bed each morning, as he anticipated what new adventures he might have in Radio Land.

>"Never give up. Never."
>"Enjoy the Moment"
>"Life is getting better all the time."
>"Radio is the intellectual's medium."
>"Smell the roses before they're gone."
>"Family is everything."
>"Take *chances*."
>"The odds are in your favor."

These were some of the phrases by which he lived. With such a positive attitude it's hard to understand why he contracted cancer. It was his usual style, of course, to live every day fully. He insisted on focusing on living instead of accepting "the inevitable." When the cancer had taken 70 percent of his eyesight, he decided to become a seminar leader. He couldn't see the expressions on the faces of his audience. He couldn't tell whether they were bored, falling asleep, or fascinated. Godfrey had a marvelous facility for telling stories that wove a mesmerizing tale so that when the audience was absolutely silent, one knew that they weren't asleep but enthralled.

Chris Lytle, whom he respected very much, told me, after Godfrey had passed on, that Godfrey had made him think new thoughts. In Godfrey's world, there can be no better gift one can give another human being. Godfrey simply saw the world a little differently than the rest of us. He never lost the wide-eyed wonder of everything—even during the seven-and-a-half years he bravely fought the cancer.

And, yes, he did finally believe that his journey with cancer was a gift. Knowing our time together here was limited truly made us appreciate every moment together. We used to joke that *we* were surprised we liked each other so much because we spent so much time together. I, Ashley, feel fortunate that I was able to be with him for 15 years. During that time he accepted me for who I was at that moment. He encouraged me to reach farther than I had ever dreamed I could. He was there to soothe my wounds when life didn't respond just the way I wanted it to. His self-effacing sense of humor always lightened my spirit.

When the cancer took him from me in March of 1995, he left behind one more precious gift. Our manuscript for this book was not finished. I languished for many months trying to avoid facing the unfinished work. As time passed, however, I began to remember and to find comfort in so many of the "Godfrey stories," which are a part of his teachings. I realized that it would not be a tribute to such an outstanding human being if I left *his* stories untold. Did Godfrey write half of this book? Yes. He wrote half of every sentence. Thank you Godfrey for being—yourself.

About the Authors

Besides being known as the Duke and Duchess of Niching, Godfrey and Ashley are radio management consultants, international seminar leaders, sales trainers, and researchers specializing in niche marketing and focus group studies.

Godfrey and Ashley have successful backgrounds in copywriting; radio, television, and print production; and media buying and media sales at the international, national, regional, and local levels. They've worked for companies like J. Walter Thompson, McCann-Erickson International, Montgomery Ward, Globetrotter Communications, and NBC. The Herwegs have owned, operated, and managed radio stations in small, medium, and large markets.

Ashley and Godfrey have also co-authored *Making More Money Selling Radio Advertising Without Numbers* and *Recruiting, Interviewing, Hiring, and Developing SUPERIOR SALESPEOPLE.*

Index

Account lists, 124–127, 242–243
Active listening, 78–79, 107
Ad agencies, 217–221, 231–233
Ad Agency Red Book, The, 221
Advertisers
 as friends, 131–132, 244–246
 interviews of, 46–53
 and niche marketing, 182–192
 as partners, 217–233
 and radio, 49–50
 win–win, 16
Advertising agencies. *See* Ad agencies
Affirmations, 129–131
Age, of customers, 73–74, 159, 163, 170
Agricultural economy, 17, 182–183
Airlines, and niches, 7, 59
Analysis, of job, 125
Appeals, advertising, 5, 10, 36–37, 93, 205
Arbitron, the, 32, 33, 88
Auto dealer, sample presentation to, 137–155
Average consumer, myth of, 37–38
Average quarter-hours (AQH), 25–26
Awareness
 of copywriters, 210
 of customers, building, 145–146
 of name, 213
 of self, 128–129
 top-of-mind, 69

Bar codes, 56, 57, 229
Best List, of advertisers, 242–243
Bias, 71–72, 79
Bicycle shop, sample presentation to, 173–180
Block groups, 32, 33
Blueprint, marketing, 173, 175–180
Books Unlimited, sample ad case, 234–236
Branding, a product, 92, 93
Budgets, 25, 54–55, 101, 232
Business cards, 123

CAB. *See* Crossley's Cooperative Analysis of Broadcasting
Cable radio, 17
Calls, sales, 36, 117, 118, 245–246
Change Quotient (C.Q.), 233
Cigarettes, advertising for, 64–65
Client Satisfaction Questionnaire, 251
Closing, 51–53, 148, 151
Coca-Cola, marketing of, 5–6, 185–187
Cold calls, 117
Commonality, 92, 211
Communication skills, 107, 245, 249, 258
Competition, 59, 209, 217
Complaints, 161–162, 241
Compliments, by salesperson, 117
Computers, 8, 34, 171–172, 183
Concepts, creative, 75, 76
Confidence, 41, 46
Confidentiality, 78–79
Conflict of interest, radio and other media, 84
Consultants
 multimedia, 88
 niche, 89
 training sales, 117–130
Contests. *See* Surveys
Control, sense of, 110, 206, 210
Cookie-cutter ads, 210
Copywriting, 193–213
 and advertising strategy, 202–205
 and awareness, 210
 and client knowledge, 206
 and commonality, 211
 and competition, 209
 and cookie-cutter ads, 210
 and focus groups, 75
 and human needs, 194–210
 and insights, 211–212
 and marketing analysis, 193–194
 and questioning, 193
 and references, 210

269

Copywriting *continued*
  and service, 211
  and surveys, 193–194
  words for, 196, 197, 199, 200, 202, 203
Core audience, 25, 36
Core demo concepts, 76
Cost
  of business. 88–89
  of focus groups, 70
  of marketing a product, 92
  per inquiry, 205
  per point (CPP), 13, 66, 222
  per prospect (CPP), 66–67
  per thousand (CPT), 222
  of promotion, 151–152
Counselors, managers as, 111–112
CPP. *See* Cost per point; Cost per prospect
Creativity, 104–105, 229–231
Credibility, 52, 192
Criticism, of ad copy, 213
Crossley's Cooperative Analysis of Broadcasting (CAB), 25
Cultural diversity 4, 5, 11, 37–38, 183
Customer Needs Analysis, 8
Customers, profile of, 47–48, 141–142

Data analysis, focus group, 83–84
Database marketing, 54–67
Databases, 27–33
  advantages of, 62–66
  building, 60–62
  and competition, 59
  and computers, 28, 56
  and demographics, 28
  formats of, 55–56
  goals of, 56
  and information retrieval, 34
  in-house, 28
  interactive, 57–59
  leasing, 66
  and margin of error, 28–29
  and researchers, 28–29
  sources of, 29–32
  and tabulations, 32
Demographic cells, 5, 27, 28, 36, 38, 62
Demographic niches, 255
Demographics
  broad, 12, 28, 36
  core, 19
  and databases, 56
  narrow, 12
  in niche presentation, 163–164
Department stores, marketing, 62
Designated market area (DMA), 33

Direct mail, 50, 190. *See also* Surveys
Direct-response advertising, 93
Diversity, cultural, 4, 5, 11, 37–38, 183
Duplication, in surveys, 171

Economics, and marketing, 64, 182–185
Economy
  agricultural, 17, 182–183
  industrial, 5, 183
  information-based, 183
  smokestack, 5, 218
Esteem, need for, 198–201
Exit surveys, 190
Expectations, clarifying, 251–254

Fears
  of advertisers, 120, 121
  choosing, 127–128
  of failure, 121
  of managers, 113
  of salespeople, 113–114, 118
  of success, 120–121
Feature selling, 96
Feedback
  negative, 115–116
  positive, 113, 244, 246. *See also* Positive reinforcement
Figures, circulation, 28
Flexibility, of salespeople, 101, 250–251
Floral niches 7–8
FM radio, lessons from, 15–16
Focus groups, 35, 68–85
  and age, 73–74
  being a client of, 77
  benefits of, 69
  bias in, 71–72, 79
  for building center, 165
  commissioning, 70–72
  composition of, 72–74
  conducting, 80–83
  data analysis, 79, 83
  goals of, 78
  guidelines, 78–79
  moderator of, 79
  outline for, 81–82
  participants, 72, 75, 78
  and power, 85
  price of, 70
  and prize selection, 73
  providing, 77
  purposes of, 68, 189–191
  report writing, 83–84
  site of, 71
  for testing, 74–76

Format, of niche presentation, 155
Fragmentation. *See* Market fragmentation
Frequent-flyer clubs, 7, 59
Friends, advertisers as, 131–132, 244–246

Gair, Sondra, 14–15
Generalship, 202–204, 213
Generosity, to customers, 248–249
Geodemographics, 13, 33
Geography, 56, 94
Giveaways, 59–60, 147, 205
Goals
 of advertising, 205–206
 of focus groups, 78
 of manager, 107
 for radio stations, 206
 for salespeople, 114
 setting, 100
Gross rating points (GRPs), 19, 20, 231
Group, myth of, 26
Groups, block, 32, 33
Growth area, of auto dealers, 145
Guidelines
 for focus groups, 78–79
 marketing, 20, 144

Hallmark Greeting Cards, 7
Hard sell, the, 96
Homogeneity, 5, 31
Hopkins, Claude, 217
Hot shops, 220, 231

Identity, testing, 74–75
Image advertising, 62–64
Independence, need for, 200–201
Industrial economy, 5, 183
Industrial Revolution, 5
Information revolution, 10, 17, 110
Information Superhighway, 34, 232
Information, sharing, 108–110
Information-based economy, 183
Insights, 101–102, 211–212
Integrated marketing, 48, 147
Interactive clubs, 57–59
Interactive databases, 57–59
Interactive radio, 15–16, 230–231, 257
Intercept surveys, 187, 191
Interviews. *See also* Surveys
 of advertiser, 46–53
 after, 83–85
 ground rules for, 80
 starting and conducting, 80–83
Intimacy, creating with client, 244–246

Lasker, Albert, 217
Lifestyle advertising, 19, 32, 36, 94–95
Listener research, 29–32
Listener survey, 31–32
Listeners
 behavior of, 26–27
 profile of, 141–142
Listening
 active, 78–79, 107
 to client, 249
 for niches, 86–96
Lists, account, 124–127, 242–243
Location, of focus groups, 71
Love, need for, 197–198
Loyalty, of customers, 248–249

Mail surveys, 30–31, 190
Managers
 and communication, 107
 as counselors, 111–112
 and fears, 113–114, 120–121
 and goals, 100, 107, 114–115
 and humanness, 112
 and mistakes, 107
 new breed of, 102–103
 and power, 102, 111
 and relations with staff, 101, 104–105, 112–113
 and self-talk, 105–106
 and sharing information, 108–110
 and success of salespeople, 106–107
Market fragmentation, 3, 5, 8, 9, 14, 87
Marketing consultants. *See* Consultants; Training
Marketing
 integrated, 48, 147
 niche, and advertisers, 182–192
 one-to-one. *See* One-to-one marketing
 presentation. *See* Presentation, marketing
 relationship, 59
 of restaurants, 60
Maslow, Abraham, 194, 197, 210
Mass appeals, failure of, 10
Mass marketing, 13–14, 37, 182, 185–188
Mass markets, and ad agencies, 218
Mass media, and the numbers, 3–4, 218
McDonald's, and niches, 6–7
Measurements, of market, 13
Media Buyer Needs Analysis, 222–224
Media buyers, 20, 222–224
Media Planner Needs Analysis, 224–229
Media planners, 20
Media, of future, 10, 257

Melting pot, 4, 87
Mistakes, and manager's attitudes, 107–108
Mobile radios, 16
Moderator, of focus groups, 79
Moms, as niches, 7
Money, for research, 25
Motivation of customers, 48, 92, 93, 142–143, 147–148, 188, 194, 203
Multimedia, and ad agencies, 218
Multimedia, and radio, 38–40

Name recognition, 213
Narrowcasting, 14, 21
National Association of Broadcasters (NAB), 149
Needs
  of consumer, 8, 26–27
  human, 194–202
  for independence, 200–201
  for love, 197–198
  of media buyer, 222–224
  of media planner, 224–229
  for nurturance, 198–199
  physical, 194–196
  for recreation, 201–202
  safety, 196–197
  self-fulfillment, 201–203
  sexual, 198–199
  survival, 194–195
Negative factors
  of job, 123–124
  of radio, 17–19
Negative feedback, 115–116
Negotiation, 101, 125–127, 221–224
Neighborhoods, as targets, 17–18
Newspapers, and advertising, 49, 157–159
Niche
  kinds of, 5–8
  listening for, 86–96
  research, and sales, 35–53. *See also* Research
  searching for, 14, 173, 188
  urban, 15
Niche managers. *See* Managers
Niche Market Analysis: Part I (NMA:I), 41–45, 48, 52, 53, 89, 93, 155, 193
Niche Marketing Analysis: Part II (NMA: II), 89–92, 93, 165, 193
Niche marketing
  and ad agencies, 220
  and advertisers, 182–192
  checklist, 20–21
  defined, 3–4
  developing programming for, 25
  and the economy, 185
  and mass marketing, 185–188
  and new approach, 89–93
  presentation. *See* Presentation, marketing
  and radio, 12–21
Nichecasters, 9, 13
NMA: I. *See* Niche Marketing Analysis: Part I
NMA: II. *See* Niche Marketing Analysis: Part II
Nurturance, need for, 198

Objections, to presentation, 146
One-to-one marketing, 13, 64, 66, 89, 156, 173, 229
Optimum Effective Scheduling (OES), 150–151
Owners, and fear of failure, 121
Ownership, of station, 103–104

Padulo, Rick, 231–232, 233
Panelists, 75–76, 84
Partners
  advertisers as, 192, 217–233
  becoming, 104–105
  clients as, 188, 215–259, 244–246
  salespeople as, 106–107, 108, 110
Partnership
  analysis of, 254
  importance of, 234–254
  losing, 251
  nurturing, 246–251
Percentages, in tabulating surveys, 171
Planners. *See* Media planners
Position, in market, 48
Positive feedback, from clients, 244, 246
Positive reinforcement, 108, 112–115, 129–131
Power, 85, 102, 111, 122
Prejudice. *See* Bias
Prequalification, of customers, 148
Presentation, marketing
  to advertiser, 182–191
  for auto dealer, 137–155
    the assignment, 137–138
    cost of promotion, 151–152
    customer profile, 141–142
    growth area, 145
    introduction, 139
    listener profile, 141
    marketing guidelines, 144
    motivation of customers, 142–143
    objections to, 146
    prequalifying customers, 148

prime market page, 140
RAB support, 144
and schedule of promotion, 149–150
summary, 152–153
for building center, 155–165
    ad plans, 164–165
    cover page, 156
    demographics, 163–164
    focus groups, 165
    format, 155
    survey results, 173–181
and newspaper ads, 156–159
Price. *See* Budgets; Cost
Prime market page, 140
Priorities, on job, 125
Prizes, 73, 147–148, 237. *See also* Giveaways
PRIZM™, 32–33
Procrastination, 113–114
Product recall, 218
Production, of ads, 229–231
Profile
    of advertiser, 48–50
    of customer, 47–48, 141–142
    of listener, 141
Programming, niche, developing, 25
Promise, in ads, 234–235, 241–242
Promotions, schedule of, 50–51
Psychographics, 13, 56

Questionnaire
    for bicycle shop, 174
    client satisfaction, 251
    for focus group, 81–82
    for furniture store, 95
    listener 31–32
    for tobacco users, 64, 65
    See also Surveys

RAB. *See* Radio Advertising Bureau
Radio
    and advertisers, 49–50
    as inventor of niche marketing, 12–21
    as new medium, 10
Radio Ad Prep Sheet (RAPS), 206–208
Radio Advertising Bureau (RAB), 52, 87, 102, 144–145, 155, 230
Radio listening, definitions of, 26
Radio Mercury Awards, 230
Radio voice, 119
RAPS. *See* Radio Ad Prep Sheet
Recreation, need for, 201
Red Book, The Ad Agency, 220–221
References, and copywriting, 210

Relationship marketing, 59
Relaxation, for salespeople, 118–119
Reliability, of client, 250–251
Reminder advertising, 62–64
Remote broadcasting, 17–18
Reports
    sharing, 108–110
    writing, 83–84
Research, 23–96
    guidelines, 78–80
    importance of, 256–257
    listener, 29–33
    and listenership, 25–34
    money for, 25
    need for, 9, 74–78
    niche, 9–10, 35–53
    successful results, 78–79
    traditional, 40
    unorthodox, 36
    up-to-the-minute, 26
    and weighting in, 28, 38
Researchers, professional, 28–29, 70–71
Restaurants, marketing, 60
Retention, of clients, 241–242
Retrieval, of information, 34
Risk, 128
Rule, the 9:30 to 4:30, 100
Rules, of ad writing, 212–213

Safety, need for, 195–196
Sales calls, 118, 245–246
Sales cycle, 51
Sales voice, 119
Salespeople
    and success, 106–107
    types of, 87–88, 101, 259
Sampling, 6, 93
Satisfaction, of clients, 243–244, 251
Scanners, 56. *See also* Bar codes
Schedule, promotional, 50–51, 149–151
Secondary markets, 40
Self-affirmations, 129–131
Self-awareness, 128–129
Self-esteem, 18
Self-fulfillment, need for, 201
Self-talk, 105, 118–119
Service, to client, 121–122, 211, 235, 248
Sexuality, needs and, 198
Share-of-advertiser, 40–41
Share-of-customer, 182, 229, 233
Share-of-market, 41, 182, 229, 233
Shoes, marketing example, 13, 185
Smart Card, 8, 230–231
Smokestack economy, 5, 218

Soft sell, the, 94–95
Software, and surveys, 171–172
Special interest groups, 5
Specialized Mobilized Radios (SMRs), 16
Station information, using, 25–27
Stations, individual, future of, 256–257
Status, need for, 198–201
Stereos, selling, 19–20
Strategy, advertising, 202–205, 224
Subconscious, programming the, 130–132
Surveys
  analysis of, 238–241
  audience of, 28
  of auto buyers, 143–144
  contest, 147, 189, 237
  and copywriting, 193–194
  duplicate responses in, 171
  exit, 190
  forms, designing, 93–94, 147–148
  intercept, 9, 187, 191
  of listeners, 31–33
  mail, 30–31, 190
  of newspaper readers, 156–157, 158–159
  percentages in, 171
  point-of-purchase, 9
  for restaurants, 60–61, 236–241
  results of, presenting, 173–181
  and software, 171–172
  tabulating, 32–33, 166–172
  telephone, 29–30, 94, 190
  universe of, 170, 238, 239
Survival, need for, 194–195

Tabulation of surveys, 32–33, 166–172
Targetability, of radio, 19
Team, building a sales, 106, 206
Technology, interactive, 16, 230–231

Telephone surveys, 29–30, 94, 190
Television
  and ad agencies, 218
  and copywriting, 210
  creative ideas, 75
  identity, 74–75
Testing, appeals, 205
Theme, of advertising, 92, 93
Time spent listening (TSL), 66
Time
  invested in clients, 242–243
  for research, 25
Title page, of presentation, 182
Top-of-mind awareness, 69
Training, sales consultants, 117–130. *See
    specific aspects*
Trust, of customers, 249–250

Universal Product Code (UPC), 57
Universe, for survey, 170, 238, 239
UPC. *See* Universal Product Code
Urban niches, 15

Value
  added, 58–59
  advertising, 35
  establishing, 9
Village Inn, The, sample ad case,
    236–241
Visualization, 118, 124
Voice, 119

Wealth, creating, 132–133
Weighting, in research, 28, 38
Wish lists, 100

Yellow pages, and advertisers, 49